Research Reports ESPRIT

Project 5345 · DIMUS · Volume 1

Edited in cooperation with
the Commission of the European Communities

S. Pfleger J. Gonçalves D. Vernon (Eds.)

Data Fusion Applications

Workshop Proceedings
Brussels, November 25, 1992

Springer-Verlag

Berlin Heidelberg New York
London Paris Tokyo
Hong Kong Barcelona
Budapest

Volume Editors

S. Pfleger
Technical University of Munich
Orleansstr. 34, D-81667 Munich, Germany

J. Gonçalves
Commission of the European Communities
Joint Research Center, I-21020 Ispra, Italy

D. Vernon
Commission of the European Communities
Rue de la Loi 200, B-1049 Brussels, Belgium

ESPRIT Project 5345 "Data Integration in Multisensor Systems (DIMUS)" belongs to the Subprogramme "Information Processing Systems and Software" of ESPRIT, the European Strategic Programme for Research und Development in Information Technology supported by the Comission of the European Communities.

The project investigates the advanced technology of multisensoring in order to obtain correct and complete scene information in monitoring and control applications with safety-critical constraints. The objective is to design and develop a system supporting a human operator in the interpretation of situations occurring in a metro (subway) station enviroment. A prototype system is being developed containing a set of tools for the integration and fusion of visual and non-visual data provided by different kind of sensors.

CR Subject Classification (1991): H.4, I.0, K.6

ISBN-13: 978-3-540-56973-2   e-ISBN-13: 978-3-642-84990-9

DOI: 10.1007/978-3-642-84990-9

Publication No. EUR 15247 EN of the
Commission of the European Communities, Dissemination of Scientific and Technical Knowhow Unit,
Directorate General Information Technologies and Industries, and Telecommunications,
Luxembourg

Typesetting: Camera-ready by authors
45/3140 – 543210 – Printed on acid-free paper

Foreword

The 1992 ESPRIT Conference and Exhibition which was held in Brussels during the week of November 23rd to 27th was remarkable for the number of technical workshops which it hosted. One of these workshops was devoted to the topic of "Data Fusion Applications" and its success is indicative of the healthy rapport which now exists between the industrial practitioners on the one hand and the research community on the other. ESPRIT - the European Strategic Programme for Research and Development in Information Technology - has played no small part in fostering this relationship. Data Fusion itself is, and will remain, an important topic in information technology. The reason is straightforward: as we attempt to expand the application domains of information technology to encompass ever more sophisticated systems and environments, the data which is derived from a single source cannot be assumed to be adequate to characterize fully the application domain. The ability to combine data derived from several sources, data fusion, to provide a coherent, informative and useful characterization then assumes a key role in the development of robust IT systems. This volume brings together a valuable collection of experiences in building systems which depend on successful data fusion and it constitutes a useful point of reference in the development of advanced IT systems in ESPRIT.

June 1993   D. E. Talbot

Preface

One is faced with many problems when designing computer-based systems which have to interpret information as it arises in natural environments. Chief among these problems is the apparent inadequacy of the available data which characterizes the application at hand, and often we can only form a partial picture of what the data is supposed to convey. On the other hand, we have the self-evident ability of humans to work extremely well in equally testing circumstances. A striking example of this ability concerns the detection of Coal Workers Pneumoconiosis (CWP), an occupational disease affecting coal miners. The early detection of CWP is effected by identifying small round opacities in a sequence of chest radiographs which have been generated at regular intervals over time. In spite of the absence of an independent reference model, CWP readers achieve remarkable performance and consistency in the classification of radiographs into one of twelve categories in the system approved by the International Labour Office, exhibiting an inter- and intra-reader standard variation which is better than one category.

The surprising factor is that CWP readers substantially disagree in pinpointing the cues, i.e. small round opacities, which form the basis for the final classification. In other words, there is strong agreement in the overall classification in spite of the fact that the same radiograph manifests different features to different readers. It can be argued that the difference in the identification of these features is a consequence of different learning sets and procedures. Whatever the case, it appears that the successful interpretation depends at least as much on the way the reader collates - or fuses - the features as much as it does upon the individual features - the data - themselves.

So how do we deal with such problems in computer-based systems? There are two obvious, and popular, solutions. The first is to improve significantly the characteristics of the sensors which are responsible for gathering the data in the first place, in the belief that with better, more reliable, and less noisy data, the problem will solve itself. While this is a valid, and frequently a necessary approach, it is by no means a panacea for these types of applications. The second solution is to try to glean as much data as possible from as many sources as possible and then to 'combine' this data in order to reconstruct and characterize the complete application domain. Unfortunately, this data fusion is not at all a trivial task and there is no single unified and proven solution which is applicable in all circumstances. Nonetheless, there are many plausible and useful approaches which can be used, and are being used, to solve particular applications in data fusion. At the 1992 ESPRIT Conference and Exhibition, the DIMUS project organized a workshop on Data Fusion Applications which emphasized the solutions to practical problems. The papers in this book describe the projects and the work presented at this workshop. A number of other papers were included later, in view of their relevance.

As one reads through this volume, one cannot but be struck by the extraordinarily wide variety of application domains in which data fusion plays a role: from the increasingly-important medical area, through industrial control, to surveillance applications in the nuclear and the transport industries. Without doubt, the scientific aspects of data fusion, i.e. the formal modelling of different information sources and their commonality which underpins successful fusion, is maturing. Equally, the validation of these theories and techniques through application has an important role to play in this maturation process. Together, they will see the development of more robust data fusion capabilities and, consequently, we will increasingly find complex problems in poorly-characterized domains yielding to solution, to the benefit of industry and the Research and Development community alike.

This book represents a unique snap-shot of the current state of play in the development and validation of technique for data fusion in a wide variety of applications, ranging from the fusion of visual and non-visual data to the integration of multi-sensor systems, and embracing many of the essential issues of, e.g., architectures, real-time responses, and safety-critical requirements. It forms a valuable reference point against which we can measure our progress in data fusion research and development, and it represents a useful foundation upon which we can build.

June 1993

S. Pfleger
J. Gonçalves
D. Vernon

Contents

Combining Two Imaging Modalities for Neuroradiological Diagnosis: 3D Representation of Cerebral Blood Vessels

Michael Bahner, Jürgen Dick, Bernd Kardatzki, Hanns Ruder, Matthias Schmidt, Arno Steitz
Theoretical Astrophysics, University of Tübingen

Carsten Bertram, Dietmar Hentschel, Thomas Hildebrand
Siemens Medical Systems, Erlangen

Eckart Hundt, Robert Kutka, Sebastian Stier
Siemens Corporate Research, München

Guido Gerig, Thomas Koller, Olaf Kübler, Gabor Szekely
Communication Technology Laboratory, Image Science Division, ETH Zürich

Abstract

Today the integration of information from different imaging modalities in medicine such as Computer Tomography or Magnetic Resonance Imaging (MRI) is left to the physician and gets little support from computers. In the case of neuroradiological diagnosis, information about cerebral blood vessels is available from 3D volume data from Magnetic Resonance Angiography (MRA) and from 2D images generated by Digital Subtraction Angiography (DSA). The DSA images have a higher resolution than MRA data, and therefore neuroradiologists are highly interested in a 3D reconstruction of cerebral blood vessels from different DSA projections. On the other hand, MRA contains important functional information, the velocity of blood flow. This paper describes work in progress to make available to the physician the full 3D information from both imaging modalities including an approach to 3D reconstruction from DSA images which makes use of the MRA data. The 3D DSA reconstruction also opens the way to an integration of information from DSA with completely different types of information, for example information on anatomical structure or soft tissue from MRI. An integral part of this work is a pilot system for clinical validation.

As a typical case the neuroradiologist is interested in a 3D representation of the blood vessels surrounding an aneurysm which would significantly support a subsequent operation. To this end MRA data and DSA images are analyzed to extract information about the blood vessels such as position, orientation, width, and branchings. These items of information are input to the 3D reconstruction based on both imaging modalities. With these results physicians are able to inspect the relevant volume in three dimensions using various visualization tools including a combined display of MRA and the 3D DSA reconstruction.

The envisaged benefits of this system are reduced patient risk due to shorter DSA examinations involving less X-ray and contrast agent and improved representations of the pathology leading to a better diagnosis and treatment.

This work is part of COVIRA (Computer Vision in Radiology), project A2003 of the AIM (Advanced Informatics in Medicine) programme of the European Commission.[1] This project started in 1992 and will last until 1994.

1. Introduction

For many decades the radiologist's working tools were conventional X-ray and film. This has changed completely during the last decade as other imaging modalities such as Computer Tomography (CT), Magnetic Resonance Imaging (MRI), or Digital Subtraction Angiography (DSA) entered clinical practice. The number of images stemming from a single patient has enormously increased and makes the use of computers indispensable. This holds especially for visualization where the display of three-dimensional data sets requires both sophisticated computer graphics and large computing resources. Far from being fully sufficient the visualization tools supplied with an imaging system at least provide a basic functionality.

But the radiologist wants more than just a separate display of the available imaging modalities. The latter provide him with different types of information about the same location in the human body, for example structural (anatomical) information from MRI and CT with functional information from Positron Emission Tomography (PET), or soft tissue anatomy from MRI with vessel anatomy from Magnetic Resonance Angiography (MRA) or DSA. This integration of information is still mainly left to the imagination of the radiologist who has to reduce a large amount of information to a concise diagnosis which he communicates to the physician responsible for the patient´s treatment. An inte-

---

[1]Participants in the COVIRA consortium are:
  Philips; Medical Systems, Best (NL) (prime contractor) and Madrid (E);
   Corporate Research, Hamburg (D)
  Siemens AG, Erlangen (D) and Munich (D)
  IBM UK Scientific Centre, Winchester (UK)
  Gregorio Maranon General Hospital, Madrid (E)
      University of Tübingen, Neuroradiology and Theoretical Astrophysics (D)
  German Cancer Research Centre, Heidelberg (D)
  University of Leuven, Neurosurgery, Radiology and Electrical Engineering (B)
  University of Utrecht, Neurosurgery and Computer Vision (NL)
  Royal Marsden Hospital/Institute of Cancer Research, Sutton (UK)
  National Hospital for Neurology and Neurosurgery, London (UK)
  Foundation of Research and Technology, Crete (GR)
  University of Sheffield (UK)
  University of Genoa (I)
  University of Aachen (D)
  University of Hamburg (D)
  Federal Institute of Technology, Zürich (CH)

grated display of different imaging modalities would simplify this process and open new ways in diagnosis and treatment, for example for operation planning. But often this integration is not a simple overlay in the event that the required information is not readily available as 3D data. Typically the radiologist has to interprete a series of slices and mentally compose a 3D image from the adjacent structures. In the case of DSA the problem is even worse because there are only a few 2D projections available, so that a surgeon has to imagine how the blood vessels visible in these projections surround a tumor visible in a series of slices from MRI.

To achieve the image integration a number of problems has to be solved. First of all a common geometry for the images has to be found (*image registration*). This can be derived from artificial markers fixed to the body of the patient or the identification of common structures in the images. Additionally geometric transformations such as interpolation will frequently be required. Since the imaging modalities produce different distortions their correction is usually necessary. A second group of problems is related to 3D visualization and integrated display, namely, apart from visualization techniques, the key problem of separating structures to be visualized from adjacent structures (*image segmentation*). Finally a seemingly minor but unfortunately a daily problem is the definition of common image data formats.

These problems are a main focus of the COVIRA project (COmputer VIsion in RAdiology) supported by the AIM program (Advanced Informatics in Medicine) of the EC. The work in progress described in this paper is part of this project and has the aim of building a clinical workstation for neuroradiological diagnosis (mainly concerned with cerebral blood vessels) which makes available the full 3D information from MRA and DSA. The DSA images exhibit smaller vessels than MRAdata, and therefore neuroradiologists are highly interested in a 3D reconstruction of cerebral blood vessels from different DSA projections. This opens the way to an integration of DSA with other imaging modalities such as MRA, MRI and CT. More specifically, an integrated system of this kind includes a 3D reconstruction from DSA projections and MRA, the matching of 3D DSA and MRA as a basis for an integrated display, and 3D visualization with a stereo screen. An integral part of the project is a clinical validation of the workstation.

The main steps of this approach are segmentation of MRA volume data which separates the blood vessel structure from the background and describes the result as a graph data structure, segmentation of DSA projections which also separates the blood vessels from the background and extracts local properties such as orientation and width, and the 3D reconstruction from the segmented MRA and DSA data with a simultaneous matching of MRA and 3D DSA. A special problem is to find a common geometry for MRA and DSA. Since a frame with markers fixed to the head and carried during MRA and DSA examinations is intolerable for the patient, a different solution has been chosen. The neuroradiologist identifies a number of anatomical landmarks in the MRA volume data and the DSA images from which an MRA-DSA coordinate transformation can be performed. However, for the 3D DSA reconstruction the imaging geometry must be known. This can in principle be read off from scales at the DSA device by the neuroradiologist during the examination, but this is not precise enough and may involve excess amount of extra work. Therefore a simple frame with markers will loosely be fixed to the patient´s head only during the DSA examination. The markers must then be identified in the DSA projections from which the imaging geometry can be calculated. This calculation, the mentioned MRA-DSA coordinate transformation and the DSA distortion correction are not described in this paper. Figure 1 summarizes the computation of the 3D representation of cerebral blood vessels from MRA and DSA.

The following sections describe the approaches to MRA segmentation, DSA segmentation, and 3D DSA reconstruction with MRA-DSA matching. Section 5 gives the design, the user interaction, and the present state of realization of the clinical workstation. The descriptions focus on the methods employed and results obtained so far. A comprehensive report on related work and the state of the art can be found in the technical annex of the COVIRA project (1991).

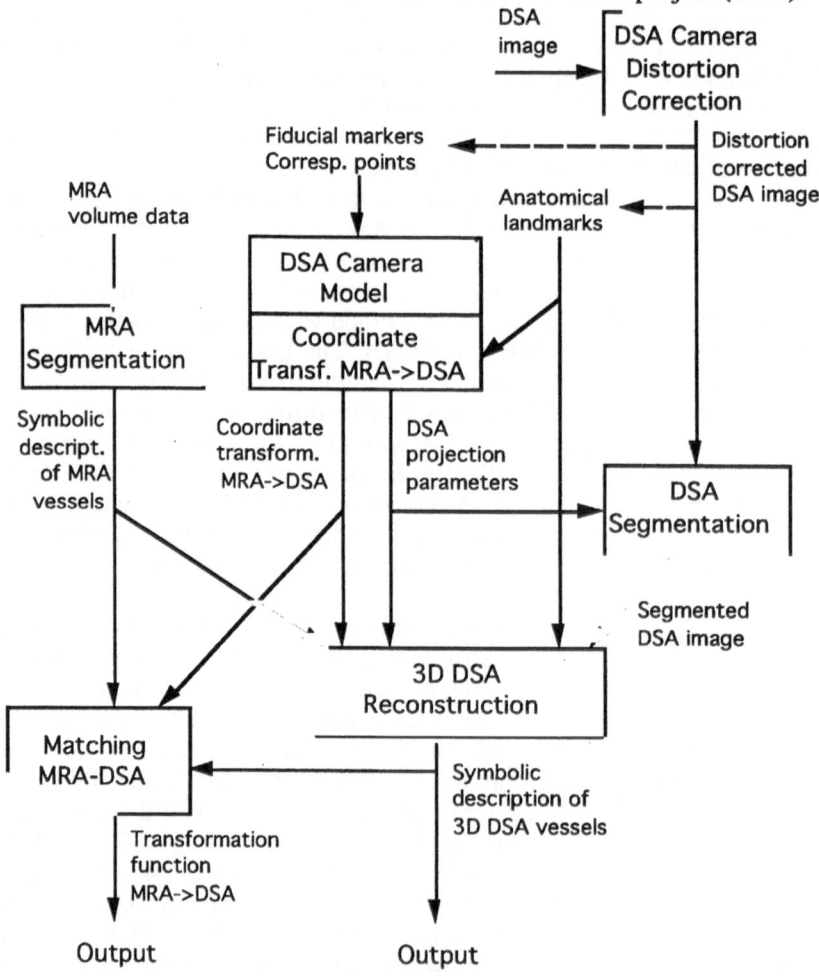

Fig.1: Computation of 3D representation of cerebral vessels.

2. MRA segmentation

2.1 3D Segmentation and symbolic description of the cerebral vascular system from MRA volume data

The basic purpose of various Magnetic Resonance Angiography (MRA) techniques is to suppress or eliminate signal from stationary tissue and to enhance the appearance of flow, and by doing so to provide optimal contrast between flowing blood and surrounding soft tissue (Fig. 2a). The segmentation of blood vessels from MRA data is an image processing task which is special in two respects. First, MRA data are three-dimensional, requiring 3-D segmentation algorithms. Second, blood vessels in MRA are represented as bright line structures. Due to magnetic field inhomogeneities, partial volume effects and noise, simple thresholding is only appropriate to segment the largest vessels. A structural segmentation approach becomes necessary to segment the complete vessel tree.

Successful segmentation into anatomically meaningful objects would create new possibilities for diagnosis, planning and intervention, but it does not provide any information characterizing the gross form or some detailed shape features. The matching with DSA projection data, however, needs topological and geometrical information describing position, width, orientation and location of branchings. Providing access to geometrical and topological properties, the binary array representation of the segmented vascular tree has to be transformed into a more appropriate data structure. Volumetric primitives of tubular structures strongly resemble the model of "generalized cylinders" with circular cross-section, suggesting the appropriateness of a skeleton or medial axis description. Furthermore, the biological variability of the vasculature favors a description based on topological features rather than on exact geometry.

We present a multi-step processing scheme including segmentation of local line features, grouping to form connected structures, skeletonization by 3-D binary thinning and graph description of object structures.

2.2 Feature extraction

The initial step to extract information from a local neighborhood at each voxel position can be described as a multiple simultaneous convolution with a set of oriented filter masks. Following Canny (1983, 1986), an optimal filter to extract lines in 2-D images comes close to the second directional derivative of a Gaussian. 3-D directional filters matching tubular structures evolve from 2-D filters by taking second derivatives of the 3-D Gaussian along two orthogonal directions. The resulting filter mask resembles a negative tubular kernel surrounded by a positive torus structure, thus supporting lines with circular cross sections. Implementing the multiple convolutions as differentiations of the image convolved with a Gaussian (linearity) results in a significant speed-up. Furthermore, one can make use of the separability and the symmetry of the Gaussian and approximate directional differentiation by discrete difference operations within 3x3x3 voxel neighborhoods.

The implementation locally calculates six second-order basis functions (Koenderink and van Doorn, 1990), from which second derivatives in arbitrary orientations can be obtained by interpolation. The second derivatives are determined either by analytically calculating the line orientation or by calculating second derivatives in a discrete number of spatial directions (regular tesselation of a sphere, e.g. vertices of a dodecahedron).

Local filtering is followed by a grouping procedure that combines local estimates to global image structures. Simple binarization of the filtered image data would result in a set of segments broken up due to noise and weak signals. Hysteresis thresholding is an alternative and works with two thresholds. An upper threshold identifies "seed" regions which are certain to be part of the sought structures. These structures are grown until a connected voxel lies below a lower threshold specifying the noise level. Figure 2 illustrates a rendered view of the segmented vessel tree.

2.3 Structural description

A possibility for generating topologically correct skeletons for filamentous tree-like structures is binary thinning (Tsai and Fu, 1981). A 3-D thinning algorithm was implemented based on three requirements: preservation of topology, maximal thinning to voxel lists and approximation of the center line. Each thinning iteration is divided into sub-cycles performing erosion of surface voxels from different orientations, mimicking parallel erosion of the whole surface. The algorithm generates curvilinear vessel "spines" represented by voxel lists (Figure 3). Parallel to the thinning operation we run a 3-D Euclidean distance transform that assigns to each voxel the distance to the nearest surface point. The distance value at the center of an object can be regarded as an approximation of the object radius. Combining the results of the two procedures, each skeleton point is attributed with a parameter expressing the local vessel radius.

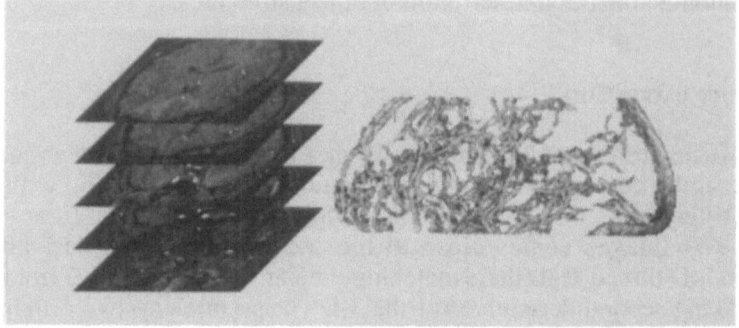

Fig. 2: Flowing blood appears bright in MRA acquisitions, as shown in the stack of axial slices (left). 3-D segmentation of the volume data extracts the cerebral vessel tree which can be visualized by 3-D rendering (right).

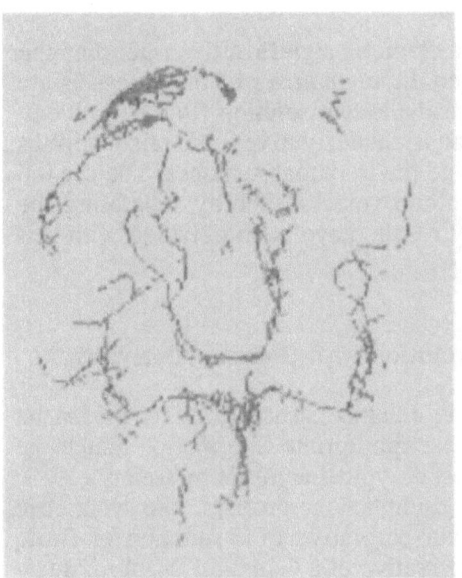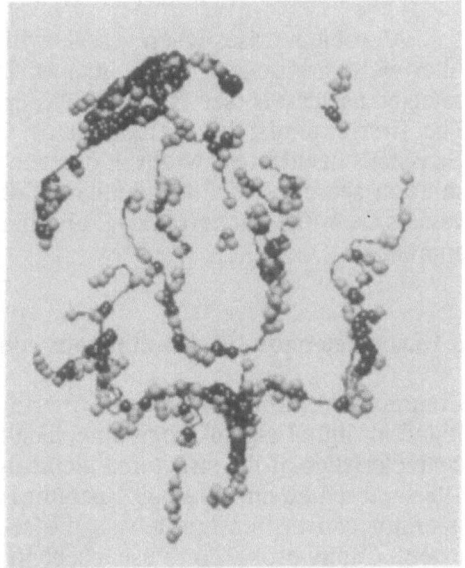

Fig. 3: Left: Result of 3-D binary thinning. The segmented vessel tree is maximally thinned to one voxel width by preserving topology and approximating center lines. Right: Graphical display of graph data structure describing the cerebral vessel-tree: End nodes and branching nodes are displayed as spheres, interconnecting vessel arcs as flexible pipes.

While retaining the complete reconstructibility, the skeleton structure serves as a data structure which can be traced by traversing lines, nodes and branchings systematically. A raster-to-vector transform "compiles" the skeleton into a graph data structure. It connects segments of pixels to sequences and assigns them to edges and vertices of a graph structure. The connectivity of voxels across faces, edges or corners is specifically considered by deleting redundant connections, resulting in a unique graph-description of the thinned structure. The vessel tree is implemented in an object-oriented manner; elements are nodes, links, points (with shape attributes) and vessel trajectories (Figure 3).

3. DSA segmentation

The second building stone - after MRA segmentation - for a full 3D representation of the cerebral vascular system is the extraction of blood vessels from 2D DSA images. To this end two approaches have been developed which are described in the following subsections. The first is based on a convolution with Gaussian derivatives of a specified width. The result are (connected) centerlines of the detected blood vessels which preserve and characterize branchings and crossings. The second approach uses a set of filters and tracks blood vessels from the detected local orientations. From the resulting orientation map, center lines and line widths can be derived.

A problem arises when vessels with a diameter significantly larger than the filter width are present in the images. Then the *boundaries* of these vessels are detected as vessels (see Figure 4). We investigate two solutions to this problem. The first extends the method using Gaussian derivatives and also applies Gaussians of different width in order to find the maximal response. The second solution searches for the boundaries of blood vessels, thereby separating the vessels from the background. Preliminary results have been obtained with this approach.

3.1 Line detection by simultaneous convolution with Gaussian derivatives

Common solutions for optimal filtering to enhance structural features can be found in digital signal processing. The most appropriate filters often match the characteristics of the structures themselves. A typical analysis by Canny (1983, 1986) presented an "optimal" operator for finding ridge profiles. The symmetric operator can be represented as the second derivative of a function of finite extent. Canny proposed to use a second derivative of a Gaussian to approximate the optimal ridge operator, suggesting that there may be an economical way of computing operators for several orientations and spatial resolutions in higher dimensions.

As in the case with ridges, the line filter is an operator composed of a detection function normal to the ridge (line) direction and a projection function parallel to it. In the case of ridges it is harder to obtain an accurate estimate of the ridge direction, especially in the presence of considerable noise. Canny (1983) suggested to simplify the problem by using several oriented masks at each point and to choose the one that best fits the ridge locally. For highly directional masks he found this approach to be an inadequate solution because it performed poorly when the ridge was highly curved. For isotropic Gaussian filter masks, however, this problem can be reduced, since the Gaussian filter is separable and derivatives can be carried out in a 3x3 pixel neighborhood.

Edge and line detection filtering produces local information about the evidence of features, but the resulting local fragments have to be grouped to global image structures. A simple, but suboptimal approach is the binarization of the filter result and a connected component labeling to uniquely assign labels to connected sets of pixels. The binarization result can be improved by a hysteresis thresholding method (Canny, 1983, 1986). Edge candidates above an upper threshold are chosen as seed elements representing enough evidence to be selected as features. A connected component labeling starts from these seeds and extends lines until its evidence falls below a lower threshold. Compared to simple thresholding, the result is less sensitive to noise, as lines with evidence between lower and upper thresholds will be discarded.

The DSA image is first convolved with a Gaussian of a specified width. In a second stage we calculate the second derivatives in four orientations. The orientation information is determined by selecting the channel with the largest output. Knowledge about the vessel polarity helps us to use constraints on the filter signal, choosing either positive or negative signals. The relative maxima are detected in the channel selected at each pixel position by looking for 1-D maxima in appropriate directions. Figure 4 shows the result of the line detection on a subregion of a DSA image.

The tests demonstrate the usefulness of hysteresis thresholding as a dynamic tool to select structures of various degree of importance. Although more robust than simple thresholding, the effect of hysteresis thresholding still depends on a careful choice of upper and lower thresholds.

A filter gives an optimal result if it matches the structure to be measured, but will fail if the structure is much smaller or larger than the filter width. As the DSA images clearly show a large variety of vessel widths, the filtering scheme has been tested at different resolutions, i.e. with different widths of the scaling parameter. As expected, the representation of large vessels has changed from parallel vessel boundaries to center lines.

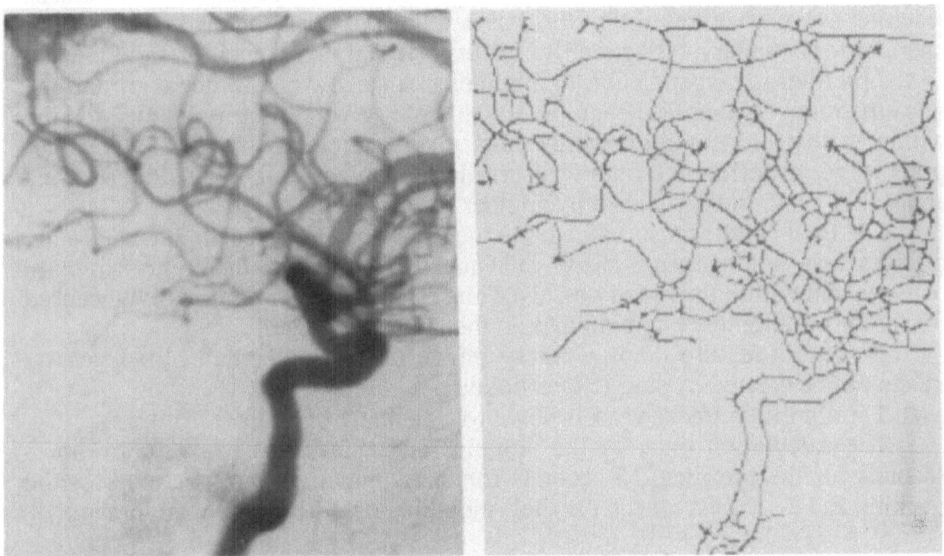

Fig. 4: Line detection procedure. Original subregion of DSA image (left). Final result after filtering and hysteresis thresholding (right).

3. 2 Line detection by convolution with line filters and tracking

The second approach to line detection uses a different set of matched filters and and additionally employs line tracking. The result of this algorithm is not only a binary line image, but also a line intensity measure, and moreover directions and diameters are calculated. Together with the center lines and branchings described above these features shall reduce the ambiguity of matching corresponding points during the 3D reconstruction. Figure 5 depicts the algorithm which consists of 4 main modules.

The first module is a star shaped prefilter which defines directions and amplitudes of prospective line structures at each point of the input image. The resulting orientation map is delivered to a connectivity module that connects neighboring pixels with similar properties by vectors. Fig. 6 shows areas which are searched for pixels with similar properties (like directions and amplitudes) Connections are drawn in both orientations of line directions, thus two vector fields are created.

The next module tracks the lines along the vectors as long as particular continuity rules are fulfilled and calculates a grey valued line intensity measure. The lines are tracked while the curvature and the branching angles (Fig. 7) are limited. The resulting line measure at a pixel equals the line amplitude at the same pixel if the search length oversteps a threshold. This definition suppresses noise structures because these consist of small amplitudes and short tracking lengths.

The final "parallel smoothing" module based on the direction field reduces resulting noise structures within the detected lines. It does not need a differential operator for boundary extraction. It uses the fact that the orientation field created by the prefilter is perpendicular to the lines in a narrow environment around the line structures. We replace each pixel value of a given input by its mean value within an environment but use only those pixels which are parallel to the center pixel. This criterion is somewhat similar to anisotropic diffusion (Perona andMalik, 1987) but is realized by another fast technique.

The final result is a smoothed line measure that enhances the intensity only of lines in a homogeneous way. That means, luminance differences remain between and within the blood vessels. Thus original thin or dark regions, caused for instance by stenoses, remain dark in the intensity image.

From this feature map, a binary image of the segmented blood vessels image can be created by simple thresholding.

The algorithm tracks vessels down to a diameter of about 2 pixels.

The calculation time for the modules on a Sun Sparc workstation are 8 seconds for the prefilter, 15 seconds for the connectivity, 10 seconds for the tracking and 9 seconds for the parallel smoothing on a 512x512 pixel image.

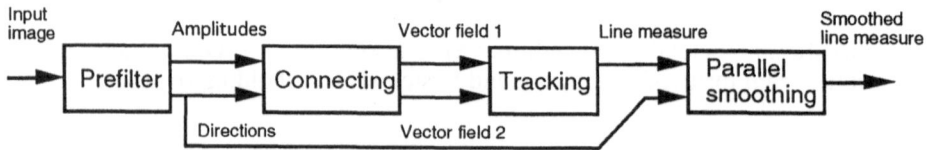

Fig. 5: Structure of the algorithm for blood vessel segmentation and extraction of an intensity measure.

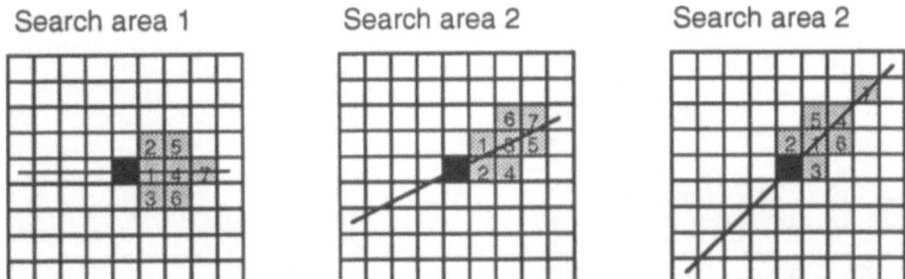

Fig. 6: For each direction, represented by a straight line, there exist search areas at both ends.

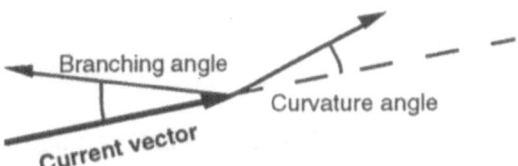

Fig. 7: Definition of the curvature and the branching angle.

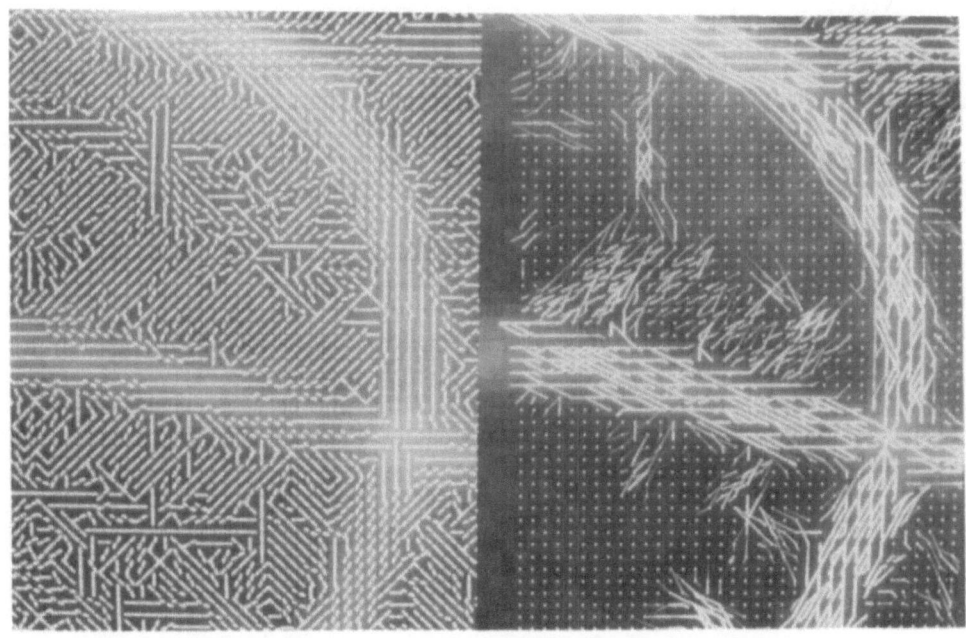

Fig. 8: Section of a cerebral DSA image. Left: the extracted directions at each pixel are visualised. Right: connectivity vectors.

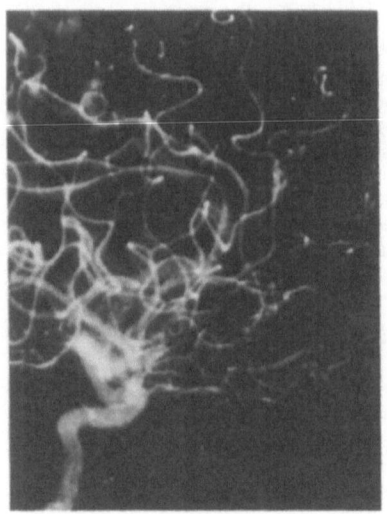 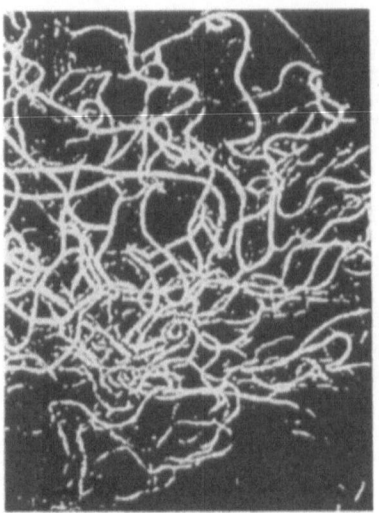

Fig. 9: Left: global view of Fig. 8. Left: original DSA image. Right: segmented image.

4. Three-dimensional DSA reconstruction and matching with MRA

This section describes the principles and the prerequisites of a new algorithm for three-dimensional reconstruction of cerebral vessel trees from few projections.

The global three-dimensional reconstruction of the brain from just a few projections is - expressed in a mathematical term - a highly underdetermined problem and not feasible in a reasonable way. If, however, the objective of the reconstruction is limited to sparse structures in the 3D volume as represented by the segmented cerebral vessel structures in DSA projection images or subregions thereof, the task is considerably simplified.

The basic idea of the new approach is to find a three-dimensional structure whose projections, if taken with the appropriate imaging parameters, match the original DSA images.

The approach heavily utilizes the fact that the cerebral vessel tree is a sparse structure in the 3D volume which can be sufficiently approximated with connected generalized cones.

Anatomical and topological properties of blood vessels are essential constraints for the reconstruction method, which can be briefly described as a combination of a model-based local fit with a global three-dimensional tracking procedure which incorporates information from segmented DSA and MRA data.

The accurate knowledge of the camera geometry of each DSA projection is substantial for the reconstruction algorithm. Since currently available DSA imaging devices do not record this information and to avoid the fixation of the patient's head relative to the DSA system, either the use of an external marker system or the interactive selection of a set of corresponding points in the projections is indispensable. From these the imaging parameters can be calculated using standard minimization methods.

The principal step in the algorithm is to initially take a single truncated cone (a base vessel element) with given start coordinates in 3D space and project it into each DSA image. A valuation of the chosen cone parameters is performed by computing the 'pixel overlap' of the cone's shade and the corresponding segmented DSA images. The combination of the normalized and weighted overlap value of each projection serves as the cost function for the quality of the fit. The reconstruction process of a single truncated cone corresponds to the minimization of this function. Each successful local reconstruction step appends the resulting cone to the vessel segment found before and yields the start coordinates for the next iteration.

Bifurcations are either detected as multiple local minima, or they are obtained directly from the segmentation methods.

Information about center lines, widths, directions, bifurcation and crossing points provided by the DSA segmentation is directly incorporated into the reconstruction procedure, primarily to get reasonable starting conditions for the minimization steps, but also to verify the reliability of the results. Segmented MRA is used to support this verification.

In case geometrically ambiguous branching topologies - e.g. the Y branching for two orthogonal projections - cannot be resolved by the method described above, each possible alternative, i.e. each local minimum, is tracked up to a decision point derived from connectivity and anatomical constraints. Branches that go astray are rejected.

The interaction to define a 3D starting point for the reconstruction consists of selecting a corresponding point in at least two DSA images and the preferred vessel direction.

To be able to incorporate information from the segmented MRA data, a common system of coordinates must be defined for MRA and DSA. Corresponding anatomical landmarks in both modalities are used to calculate the parameters to transform the MRA data into the DSA coordinates.

So far this algorithm has achieved the reconstruction of projections generated from synthetic tube structures. The results indicate that the described approach is feasible also for real data.

Additionally the combination of segmented MRA and reconstructed 3D DSA leads to a new approach to distortion correction of MRA images. By identifying several corresponding anatomical landmarks - e.g. bifurcations - in both the MRA and the 3D DSA vessel tree, points are gained which are useable for performing a correction of distortions in the MRA data caused by flow phenomena and magnetic field inhomogeneities.

5. Clinical workstation

One of the major goals of the COVIRA project is to gain experience with new image processing techniques and to assess their usefulnes in clinical practice. Therefore an integral part of the work described in this paper is the development of the software package NEUROVISION which will also incorporate modules from other work packages of the COVIRA project. Clinical validation is done at the University Hospital of Tübingen, where a first clinical prototype is in use since January 1993.

The platform for NEUROVISION is a commercially available general purpose workstation, running the UNIX (T) operating system. Two monitors are used for viewing and display, one of them a special stereo monitor. An Ethernet connection allows an easy and fast transfer of the acquired patient data from the MR and DSA imaging devices.

A principal guideline of our software design has been to keep the architecture modular. This is necessary to allow an easy integration of new software delivered during the project. Our most important task has been to build a simple, consistent and efficient user interface. We achieve this by using a graphical user interface which is based on standard techniques like windows, buttons and mouse handling. Furthermore, we use the direct manipulation technique: a cut plane, for example, is interactively specified by moving and/or rotating a reference line representing the cut.

The software currently consists of the following modules: image transfer, image database, image display, 2D image processing, 3D volume data processing, distortion correction, image registration and fusion.

The image transfer module reads in the image datasets sent by the imaging devices, converts them if necessary to the ACR/NEMA format and stores them in the local database on the workstation.

The image database provides access to all images of one patient, divided into hospitalizations and examinations, and stores calculated images. The base unit for organizing the images is a folder.

The image display module allows the user to arrange the number and size of images on the screen as desired. Furthermore, it is possible to view DSA images acquired under stereoscopic conditions on the stereo monitor. This feature allows the radiologist to obtain a stereo impression of the blood vessels.

2D image processing includes tools for zooming, paning, or analyzing images.

The module for 3D volume data processing provides some basic functionality for the manipulation of volume datasets. A first step is multiplanar reformatting (MPR), the calculation of arbitrary orthogonal, oblique or double oblique cuts. While vessels are the objects of interest for our application, another important 3D operation is the maximum intensity projection (MIP), where parallel rays are cast from arbitrary directions through the volume and the highest intensity for each ray is displayed. For MRA this is the preferred method for the display of vessels. The computation and visualization of MIPs under stereoscopic conditions gives an improved impression of the vessels' spatial relation.

The distortion correction module provides methods for the quantification and correction of geometrical distortions for DSA and MR. MR distortions are caused by magnetic field inhomogeneities and patient-related susceptibility changes whereas DSA distortions are due to the image intensifier. Based on phantom measurements the scanner related distortion is determined, so that in a second step the patient data can be corrected.

The registration module determines the transformation equations between different datasets. In the case of MR and DSA fusion two methods are possible. First, the 2D projection of the MR data according to the DSA geometry and second, the fusion of the 3D MR data with the reconstructed 3D DSA data set. While the second method is one of the main topics but not yet completed, the first method is already implemented. The 2D projection of the MR data according to the DSA geometry is done in two steps. First the 3D imaging geometry of the DSA device has to be determined using a frame. A tool assists the user in the identification and assignment of some frame markers. Based on the known frame geometry the camera model is then calculated. The second step is the determination of anatomical landmarks visible both in the MRA and DSA dataset. A tool supports this task by utilizing the MIP functionality, which is appropiate for a fast visualization and identification of vessels in the MRA dataset. Determining a vessel structure in MRA means to fix a point in 3D. For that the user identifies a landmark in one of the three orthogonal MIPs. The point is traced back in the 3D dataset so that the maximum intensity along the ray gives the third coordinate. Both the ray and the point with the maximum intensity are displayed in the other two MIPs for a manual correction if necessary. The structure identified in MRA is then indicated in the DSA image. Using these assignments and the camera model from the first step the transformation of the MRA dataset into the DSA geometry can be done.

The image fusion module allows the display of previously registered data sets. Two modes are available for that: with common points an indicated point in the MRA dataset is marked in the corresponding DSA projections. This allows the physician to compare the same location separately in the two modalities. Overlay projects the MR data set according to the DSA camera model, the resulting 2D image is merged with the DSA image.

6. Conclusion

This paper describes an approach to the integration of information from different imaging modalities in the field of neuroradiology. The approach comprises the extraction of the blood vessel structure from MRA volume data and 2D DSA images which is input to the 3D DSA reconstruction and matching of MRA with 3D DSA. This process is realized on a clinical workstation directly connected to MRA and DSA systems. This workstation has recently been installed at the Department of Neuroradiology, University Hospital of Tübingen, next to the DSA examination room, and will undergo a clinical validation.

A major benefit of this system will be the reduction of the patient's risk during DSA examinations which have a certain inherent risk so that they are required to be complete, fast, and safe. A 3D DSA reconstruction will reduce the number of DSA images taken during an examination because it will no longer be necessary for the neuroradiologist to search for projections showing the relevant topographic information. Therefore the examination will not only be faster but also less X-ray and contrast agent will be used. A second benefit will be a more precise delineation of the pathology and its surrounding vessels with the possibility to obtain arbitrary topographic information by inspecting this volume from appropriate directions. Simultaneous display of MRA and DSA enables the neuroradiologist to combine characteristic information from each modality, i.e. flow information from MRA and precise geometry from DSA. In the case of an overlay of 3D DSA with CT or MRI, completely different types of information - blood vessel structure with anatomical structures and soft tissue - will be combined. A final benefit of this approach is that it will be easier for the neuroradiologist to communicate the diagnosis to the physician subsequently treating the patient, e.g. a neurosurgeon.

References

1. Canny J.F. Finding edges and lines in images, Technical Report 720, MIT Artificial Intelligence Laboratory, Dept. of Electrical Engineering and Computer Science, Cambridge, MA, 1983.
2. Canny J.F. A computational approach to edge detection, IEEE Transactions on Pattern Analysis and Machine Intelligence, vol.8(6), pp. 679-698, 1986.
3. Computer Vision in Radiology (COVIRA), project A2003 of the AIM (Advanced Informatics in Medicine) programme of the European Commission, Technical Annex, pages A1-1 - A5-6, Brussels, 1991.
4. Koenderink J.J. and van Dorn A.J.. Receptive field families (generic neighborhood operators), Biol. Cybern., vol. 63, pp. 291-297, 1990.
5. Tsao Y.F. and Fu K.S. A parallel thinning algorithm for 3D pictures, Computer Graphics and Image Processing, vol.17, pp. 315-331, 1981.
6. Perona P. and Malik J. Scale space and edge detection using anisotropic diffusion, Proceedings of IEEE Workshop on Computer Vision, Miami, pp. 16-22, November 1987.

Hybrid Inference Components for Monitoring of Artificial Respiration

K. Gärtner, S. Fuchs, H. Jauch
Institute of Artificial Intelligence
Dresden University of Technology

Abstract

Medical expert systems are developed for supporting the complicated decision finding processes of the physicians. Generally, the decision finding in a clinical therapeutic process contains a complex data fusion problem. The problem is analysed and the conclusions are used for the development of medical expert systems.

The indentified pattern of problem solving behavior of the experts allows the structuring of the knowledge in sections. Different data structures need different representations and inference strategies. The higher organized expense for the data fusion is solved by implementation of a blackboard structure.

Finally, hybrid inference components for monitoring of artificial ventilation are proposed.

1. Introduction

The development of medical expert systems make endeavour to support the complicated decision finding processes, required from physicians in diagnosis and therapy (see figure 1).

Generally, these processes are complex data fusion problems. Each part of a clinical therapeutic process may be an information source as well as an information destination.

The conventional procedure consists of the relations, which are presented in figure 1. These relations correspond to destinctive partial processes named in table 1 (upper left part). By the integration of computer the architecture of the procedure could be simplified in the sense that all parts only exchange information via the computer (see table 1, heary bounded part). The control and the adjustment of the devices may be made by the computer and additional the data fusion and decision making also may be done. But because of the patient's safety and the responsibility of all therapeutic reactions the computer has to be controlled by the physicians and therefore he must get also information from

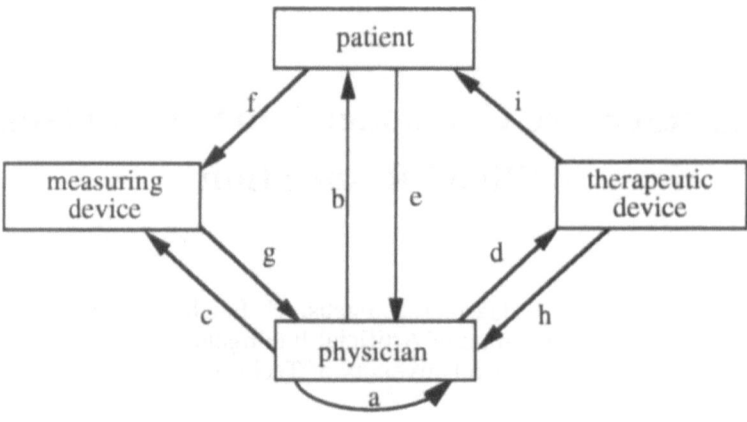

Conventional Procedure

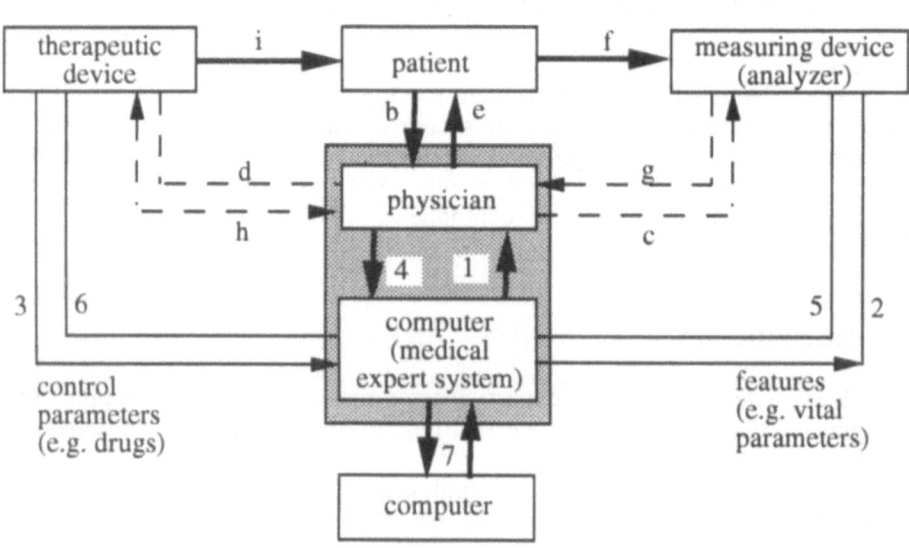

Fig. 1: Block sheme of a clinic therapeutic process with interactive decision finding

the measuring device and the patient directly. Typically, in this scheme the job of the computer is an expert system.

For the development of such an expert system for clinical therapeutic processes it is necessary, that all data from each information source are processing by the computer. The processing capabilities of computers make possible the widespread dissemination of medical expertise.

Access to large amounts of information about different treatment methods are likely to improve the ability of physicians to manage patients with complex problems. Diagnosis finding and therapy planning are important areas for the development of computer-based decision aids.

Although, recently we have been collected many experience with computer-based techniques to assist the management of difficult therapeutic problems, most applications describe the treatment of patients for whom the course of the disease and the response to therapy are relatively stereotypic.

destination sources	physician	patient	measuring device	therapeutic device	computer
physician	a consultation	b examination	c adjustment of measuring ranges	d adjustment by hand	1 simulation, consultation
patient	e examination	----------	f measurement	----------	----------
measuring device	g condition information	----------	----------	----------	2 automatic measurement
therapeutic device	h adjustment control	i therapy	----------	----------	3 automatic control
computer	4 proposal	----------	5 adjustment of measuring ranges	6 (adjustment)	7 simulation, consultation in databases

Tab.1: Information sources and destination by medical therapeutic processes

For example, knowledge in the ONCOCIN system is largely composed of algorithmic rules derived from making large protocols, that anticipate potential problems with therapy and list guidelines for appropriate treatment responses [1].

These rules often do not contain the expert knowledge necessary to provide high quality advice for those patients which do not respond to conventional therapy or which experience unusual complications [2].

Nevertheless, expert physicians can devise carefully reasoned treatments for such difficult cases.

2. Characterization of therapeutic processes

The complexity of the problems in the medical therapy is the result of the following features, which distinguish these problems from non-medical problems:
1. No unique mapping between measured quantities and controlled quantities (For instance: different modifications of ventilator parameters can result in same success concerning change of the vital parameters).
2. Uncertainty and vagueness by the valuation of the actual patient condition (There are not often diagnosed neither basic disease nor pulmonal condition of the patient.).
3. Unknown, possibly incidental influences on the treatment (The result of the treatment can not often be foreseen, for instance the effect of the therapy by drugs).
4. Compromises with the choise of the therapy:
    a) trade-off between the negative effects of the individual ventilator parameters (e. g. the positive endexpiratory pressure or the inspired fraction of oxygen)
    b) permanent consideration of the effects to other organs.

3. Problem solution

The analysis of the therapeutic process by clinical-experimental examinations with following discussion is very important for the development of expert systems for the medical therapy.

3.1 Problem solution for therapeutic processes

Four aspects of the problem solution for the therapeutic processes are here presented:

A first aspect: By the decision finding in the medical therapy a consistent pattern of problem solving behavior of the experts was indentified:
• Development of a set of possible plans that are reasonable to administer.
• Estimation of the possible consequences of each of these plans.
• Valuation of conformity between predicted consequences of each plan and the treatment goals.
This process conception should be subjected to the construction of computer-based therapy decision finding systems. Artificial intelligence techniques, simulation and decision finding theory should be integrated at the best suited stage of the planning process. Examinations showed, that each method alone is not efficient enough for the solution of therapeutic problems which have properties mentioned in section 2.

A second aspect: Modern therapeutic consultation and expert systems have got a high degree of specialization. With increasing volume of the knowledge base the acquisition becomes more and more difficult and the expense by computation raises [3]. For that reason it is better, that the knowledge is structured in sections, only the needed section is always actualized. By that the complete solution of a complex problem is configurated by a number of specialized, closed and re-used components. The knowledge bases can be installed on one or several computers. The higher organizing expense can be handled by implementation of a global data structure, called blackboard.

A third aspect: The data structures of different information sources are not uniform. Different data need different representations (different models) for the computer-based processing. Hybrid inference components for an expert system in the medical therapy originate obligatory and consequent from different data representations.

A fourth aspect: The data from the different information sources are partially fuzzy and incomplete. Table 2 describes the characteristics of the data. Fuzzy and incomplete data require the utilization of fuzzy recognition methods.

information source	data properties	
	structure	condition
measuring device	numerical	vague and incomplete
physician	verbal in rules	uncertain and incomplete
therapeutic device	numerical	vague

Tab. 2: Data properties of the information sources

3.2 Problem solution for optimization of artificial ventilation

The medical consultation system is called IBEUS, a german abbreviation for Intensive-therapeutic-ventilation-decision-support-system.

With IBEUS we want to contribute to optimization of artificial ventilation by means of continual tuning the important parameters of patients and devices. The aim is an adequate supply of the patient with oxygen of little pressure and incidental consequences to other organs.

By reason of the medical specialities we decided for a concept, making inferences from

1. knowledge, which is contained in the measured quantities of the patient (information source : measuring device) or in the quantities controlled by the physician (information source : therapeutic device).
2. knowledge, which is reflected by experience and literature (medical expertise, information source : physician)

For choosing off the suitable inference strategy a structured knowledge representation is demanded which is in correspondence with the medical decision process.

The structuring of the medical decision process may be provided by the differential diagnosis aspects.

- The first step in the differential diagnosis procedure is the actual indication.
- The second step encloses search for the important features for the description of the vital patient condition (PEEP, PaO_2, $PaCO_2$, and so on)
- By interpretation of these features the physician decides, in which therapeutical group the treatment is classified. With the features and the therapeutic group a suitable ventilation concept (proposal about the change of the ventilatorparameters) can be estimated.
- Lastly the quantity must be estimated and the conditions (e.g. change of the parameter by actual indication?) and limiting values have to be controlled.

The last two steps are hardly connected. The result is the confirmed proposal for the ventilator parameters.

The process based on these four aspects is represented by the following graph.

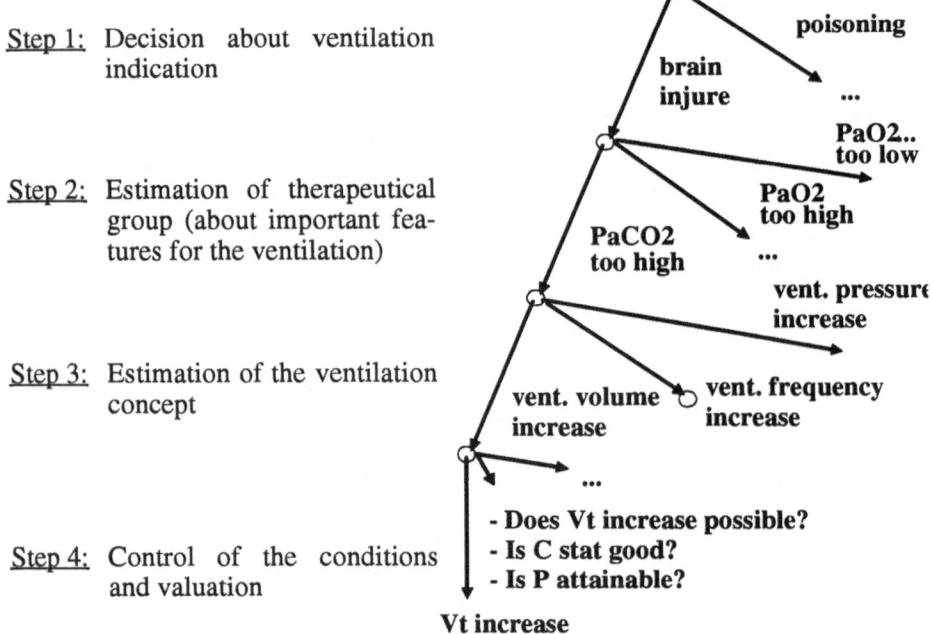

Step 1: Decision about ventilation indication

Step 2: Estimation of therapeutical group (about important features for the ventilation)

Step 3: Estimation of the ventilation concept

Step 4: Control of the conditions and valuation

Fig. 2: Structuring of the medicial decision process by optimization of artificial ventilation

Figure 2 is only a simplified model of the decision finding process by the optimization of artificial ventilation. Now the conception of the decision support system shall be described. The base of the IBEUS conception is a blackboard architecture. The knowledge is divided with respect to knowledge sources accomplishing the partial problems. The solving knowledge sources at present state are shown in figure 3.

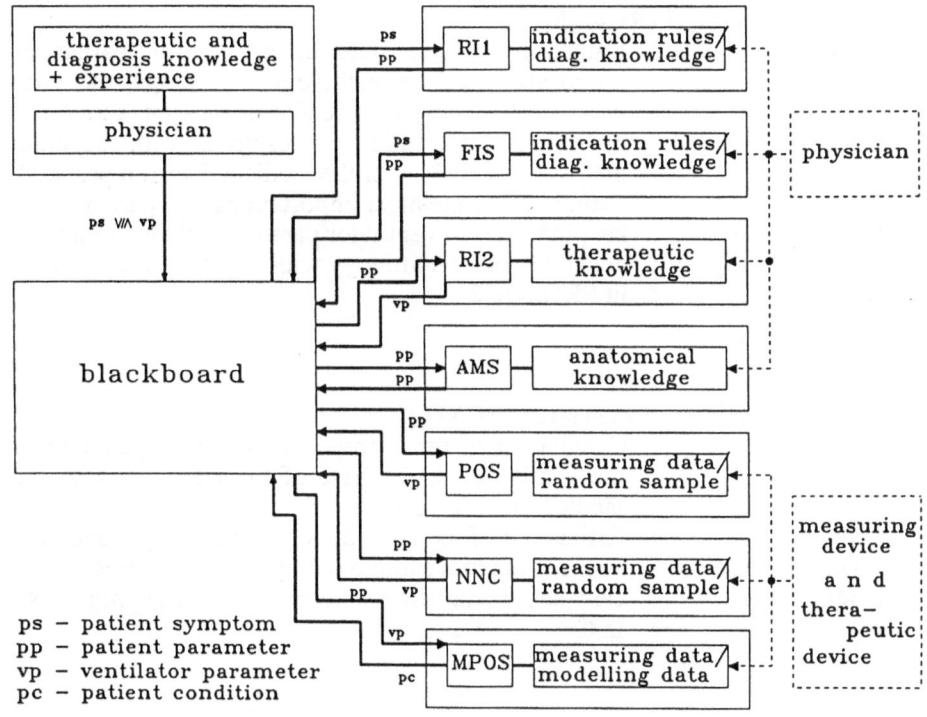

Fig. 3: Knowledge sources of IBEUS

The system function is accomplished by the physician and seven partial processes working on the blackboard. They are characterized in the following passage.

1. Rule interpreter I (RI 1):

- Goal : Estimation of the most important patient parameters for the decision (step 1)
- Condition : Demanding the actual pathological findings and important additional information about the patient's condition from the physician
- Knowledge base : keeping rules for the indication and illness
- Inference strategy : case-data-driven and forward chaining
- Realizing : in PROLOG2

2. Fuzzy inference system (FIS):

- Goal : Estimation of the most important patient parameters for the decision (step 2)
- Condition : Demanding the actual pathological findings and important additional information about the patient's condition from the physician
- Knowledge base : keeping rules for the indication and illness
- Inference strategy : simple approximate reasoning or plausible reasoning
- Realizing : in C

3. Rule interpreter II (RI 2) [5]:

- Goal : Valuation of the determined control parameter proposals based on pathological findings, the vital parameter values, the trend information and other additional information or estimation of separate proposals (step 2 - 4 of the medical decision process)
- Condition : Estimation of the most important patient parameters
- Knowledge base : keeping rules to ventilator parameter changes and
- Inference strategy : generate-and-test-strategy with differential diagnosis
- Realizing : in PROLOG2

4. Prognosis optimization system (POS) [4]:

- Goal : Calculation of three proposals for the adjustment of the ventilator on the base of the (step 2 - 4 of the medical decision process)
- Condition : Estimation of the most important patient parameters
- Knowledge base : keeping random sample with patient's features
- Inference strategy : case-based (statistical numerical model to prognosis
- Realizing : in C

5. Neural network component (NNC):

- Goal : Calculation of a proposals for the adjustment of the ventilator on the base of the determined actual parameters (step 2 - 4 of the medical decision process)
- Condition : Estimation of the most important patient parameters
- Knowledge base : keeping random sample with patient's features
- Inference strategy : back-propagation
- Realizing : in C

NAME	DESCRIPTION
AMS	Respiratory mechanics simulation
FIS	Fuzzy inference system
MPOS	Modified prognosis optimization system
NNC	Neural network component
POS	Prognosis optimization system
RI 1	Rule interpreter I
RI 2	Rule interpreter II

Tab. 3: Process working on the blackboard

6. Modified prognosis optimization system (MPOS) [6]:

- Goal   : Estimation of a prognosis for a description of patient's condition one hour after change of the ventilator parameters
- Condition   : medical proposal for the ventilator parameters
- Knowledge base   : keeping random sample with patient's features and knowledge for the condition description over prognosis values
- Inference strategy : case-based over statistical numerical model to prognosis optimization
- Realizing   : in C

7. Respiratory mechanics simulation (AMS) [7]:

- Goal   : Estimation of conclusion over the system behaviour (resiratory mechanics) over the model behaviour
- Knowledge base   : keeping differential equations to modelling of respiratory mechanics
- Inference strategy : model-based over differential equations
- Realizing   : in C

The structure of the blackboard reflects approximative the medical decision process.
In this system the blackboard is the base of data fusion.

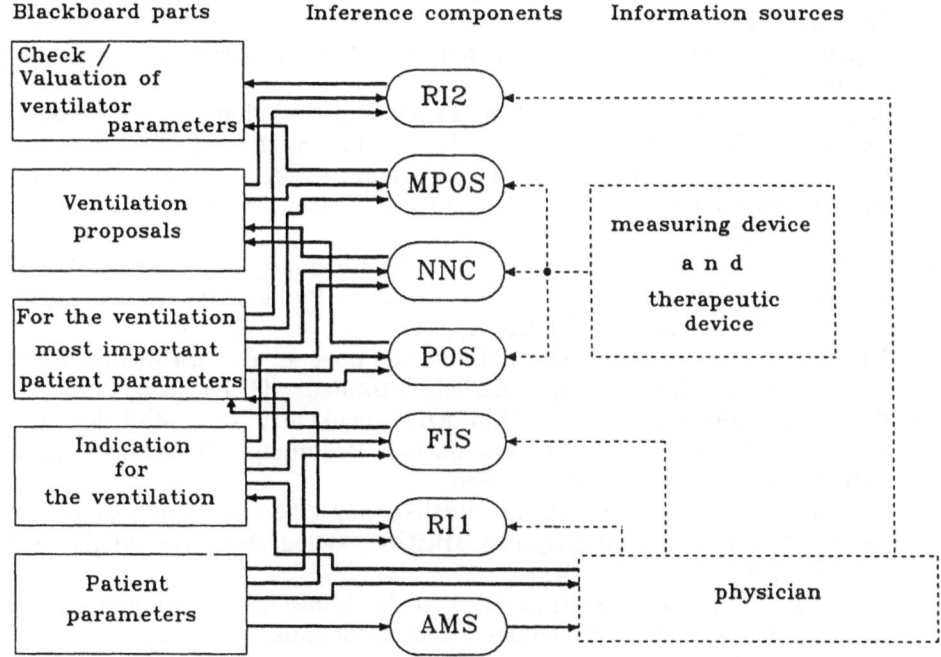

Fig. 4: Structure of the blackboard for IBEUS

For the physicians the following possibilities and advantages arise in the work with IBEUS:

- Calculation of unvalued (RI1 - POS; FIS - POS, RI1 - NNC; FIS - NNC; RI1 - RI2; FIS - RI2)or of valued proposals (RI1 - POS - RI2; RI1 - NNC - RI2; FIS - POS - RI2; FIS - NNC - RI2) to the optimization of the ventilator parameters for the actual ventilation case (different ways to data fusion).
- Test of selected ventilator parameters patterns at actual or fictive patients by calculation of the possible patient condition by means of prognosis value (RI1 - MPOS; FIS - MPOS).
- Information about the system behaviour (only AMS)
- Justification of therapeutic solutions (explanation components by RI1, RI2)
- Possibilities to knowledge acquisition
    a) statistical learning component to renovation of the model parameters by enlarged random sample [8]
    b) knowledge editor to enlargement of the rule-based knowledge base [9]
    c) direct knowledge aquisition by inductive reasoning over the random sample [10].

References

1. Langlotz C.P., Fagan L.M., Tu S.W., Sikic B.I., Shortliffe E.H., "A Therapy Planning Architecture that combines Decision Theory and Artificial Intelligence Techniques", In: Computers and Biomedical Research 20, S. 279-303, 1987.
2. Puppe F., "Problemlösungsmethoden mit Expertensystemen", Springer Berlin, 1987.
3. Fagan L.M., "Representing Time-Depending Relations in a Medical Setting", USA, Standford University, Doctoral thesis (Ph.D.), 1980.
4. Gärtner K., "Problemlösung für die Regelung der maschinellen Beatmung auf Intensivstationen unter Nutzung der rechnergestützten Entscheidungsfindung", Diss., IH Dresden, 1984.
5. Schreiber Th., "Beratungssystem für die Beatmungsüberwachung unter Nutzung statistischer und wissensbasierter Problemlösungsmethoden", Diss., Fakultät Informatik, TU Dresden, 1990.
6. Jauch H., "Konzipierung einer Inferenzkomponente für das medizinische Beratungssystem IBEUS", Diplomarbeit, Informatik, TU Dresden, 1992.
7. Kaiser S., Morgenstern U., "Einsatzmöglichkeiten von Modellen der Ventilationsmechanik bei der maschinellen Beatmung", Wissenschaftliche Beiträge der IH Dresden, Heft 5, 1986.
8. Grohmann U., "Konzipierung und Realisierung einer Lernkomponente für das medizinische Beratungssystem IBEUS", Belegarbeit, Informatik, TU Dresden, 1992.
9. Pöthig A., "Wissensbasis-prototyping in der klinischen Praxis anhand der ARDS-Grundwissensbasis", Diplomarbeit, Informatik, TU Dresden, 1990.
10. Brandt A., "Auswertung metrischer Stichprobendaten unter Anwendung clusteranalytischer Verfahren", Diplomarbeit, Informatik, TU Dresden, 1992.

Information Fusion in Monitoring Applications using the Category Model

Wolfgang Steinke
Siemens AG, SI E SW 1, P.O. Box 1661
D-8044 Unterschleissheim, Germany

Abstract

In many application areas similar monitoring tasks can be identified. The fundamental monitoring tasks are reported based on a technical domain analysis within different application areas. Common to most of the monitoring tasks are the recognition, surveillance, and control of objects, states and/or processes within a specific area.

Important for performing those basic tasks is the collection and processing of data coming from different sources. A typical characteristic of data acquired from different sources is the different type of data and the varying level of information. The process of combining data concerning a specific goal stated by a monitoring task is often called *Data Fusion*.

Here the more general term *Information Fusion* will be introduced to describe the process of condensing data on varying level of information. The process of Information Fusion is based on an approach for a goal-oriented handling of information, the Category Model, which serves as the representational framework for the fundamental monitoring tasks.

Several results of this work have been performed within a cooperation with the ESPRIT-Project 5345 DIMUS (Data Integration in Multisensor Systems).

1. Introduction

There are many application areas in which monitoring and control tasks become more and more important. While in former times these tasks have been performed by human operators, today computer-based systems fulfil these tasks. What are the reasons for this process?

In many application areas, in which operators are responsible for conducting monitoring tasks, the responsibility is stressed to maximum load. The complexity of the environments to be monitored and the continuously increasing amount of data to become controlled, increase the need of computer support for the human operator. The need for the operator support appears mainly in those application areas, in which monitoring, control and decision tasks have to be performed, and where fast, safe and reliable reactions of the operator are required. The advantages of automated monitoring tasks can relief the operator load and allow the operator to concentrate on the supervision tasks.

The structure of monitoring systems, supporting an operator, depend basically on the environment to be monitored and the interface between environment and the application.

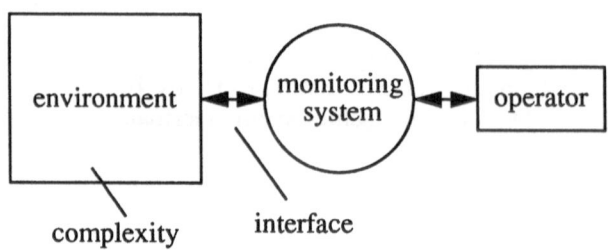

Fig. 1: The basic situation

For the environment point of view, the complexity of the system or process to be monitored decides which functions become automated and which functions remain in the responsibility of the operator. Deciding, whether a function becomes automated or not, depends on several different factors, i.e. the transparency of the process to be monitored, the availability of methods and techniques for handling the complexity of the process.

Important for the handling of the input information within a monitoring system is the kind of interface between the environment and the application. A broad range of sources of information do exist, leading from the world of sensors to natural language as an input. While basic types of sensors produce binary signals, the more sophisticated concept of the logical sensor [1] abstracts from the characteristic physical features of the sensors. The term *human sensor* is placed at the other edge of the scale of information sources and stands for the broad spectrum of symbolic input.

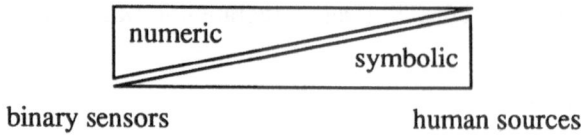

Fig. 2: Degree of abstraction

Not only the different type of information depending on its sources is responsible for the need of automation. Also the increasing quantities of data to be handled by the operator lead to a situation, where a reliable monitoring is not longer guaranteed.

The forces which influence the basic situation of monitoring tasks is given by the environmental complexity, the quality of data and/or information and the quantity the data is made available. The strength of these forces distinguish the limits for the operator and the monitoring system.

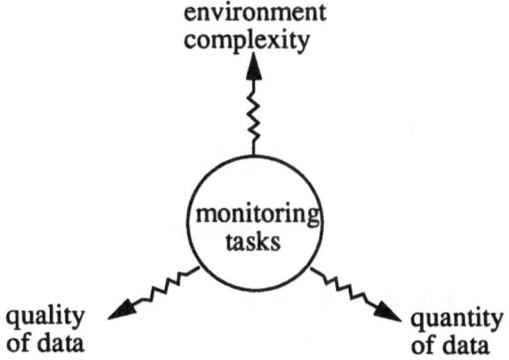

Fig. 3: The monitoring tasks influenced by forces

Driven by these forces, the motivation for the development of a *generalized model of monitoring tasks* arises, where the process of handling inputs from different sources and of different types - known as *(data) fusion* process [2,3,4]- plays a specific role. The term *information fusion* is used in order to describe the abstract process for collecting and condensing data to more monitoring-goal-oriented items.

The first step towards a generalized monitoring model is the analysis of systems used in different monitoring application areas. Sometimes, this process is called *(technical) domain analysis*. Functional similarities of the systems are being analyzed and lead to a number of fundamental monitoring tasks. These monitoring tasks represent the functions which are common to most of the analyzed systems. What application areas have been analyzed and what the fundamental tasks are, is subject of the next section.

The second step to a *generalized monitoring model* is, to answer the question: What are the goals which shall be reached by those fundamental monitoring tasks?

One answer to that question is given by the **Category Model**, which serves as a representational framework for the fundamental monitoring tasks. Section 3 highlights the concept of the category model and presents the level of information fusion based on the category model.

2. Application Areas and Fundamental Tasks

The analyzed monitoring applications are coming out of different domains. In these domains most aspects of the basic situation described above do exist. An operator is monitoring a complex environment using several different input sources for reaching specific goals.

For the analysis of monitoring applications the domains of indication & warning systems, automation, tactical situation assessment and air traffic management have been selected (see figure 4).

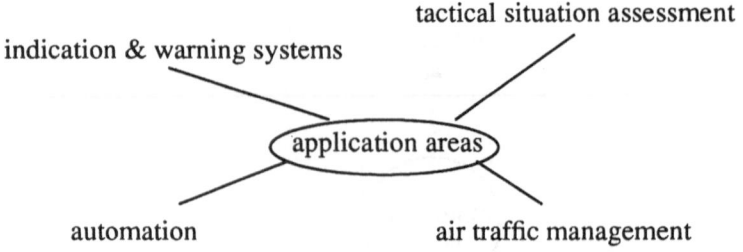

Fig. 4: Selected application areas

Within the domain of indication & warning systems one will find systems where characteristic events or situations are indicated and alarms have to be handled, i.e. the surveillance of subway stations [11,12].

In the automation domain offers applications leading from process control tasks to autonomous robots. Herein, the operator has a very varying responsibility [9,10].

The objective of tactical situation assessment is to acquire and combine data from sensors as well as messages from human observers to generate a tactical picture of the environment [4,5,6]. Based on this tactical picture military decision are prepared.

The air traffic management domain covers the requirements for managing the traffic in the air as well as on the ground. For handling the complex air traffic environment, international laws and rules regulate the procedures of the air traffic[7,8].

Definitely, the selection is not complete, but often, the requirements of systems being not part of the selected domains, are common to the requirements of the analyzed set of applications. Nevertheless, the selected applications give a good overview, what the application areas are in which information fusion techniques are used for monitoring tasks.

Analyzing those applications, one will recognize, that often the basic goals of applications are similar. For example, the monitoring and determination of the position of an object is the goal of a lot of monitoring systems. This task can be identified in the air traffic management domain as well as in military applications or indication & warning systems. Figure 5 shows three monitoring tasks, where the common goal is the position identification of an object.

The identification of the position of (a part of) troops and the monitoring of their movements is a well known goal of tactical situation assessment systems within the military domain.

The monitoring of aeroplanes is subject of ground movement control systems in the air traffic management domain. Herein, the position of an aeroplane has to be identified for a safe taxiing of the aeroplane on the airport.

Within an subway surveillance system [12], the position of persons and objects has to be monitored, to detect i.e. the violence of restricted areas like the railway or the tunnel entrance.

monitoring tasks: position identification			
application area		object	goals
military	tac. situation assessment	(parts of) troops	identification in which direction a troop moves
air traffic management	ground move-ment control	aeroplane	tracking of the aeroplane on taxi routes
indication & warning	subway surveillance	persons, ...	identification of objects in restricted areas

Fig. 5: Similarities in monitoring tasks

Common to all those tasks is the goal of position identification of an object. But not only the task is similar: to reach the goal of position identification, each application has to collect, evaluate, combine data of different sources and of different type. This process of information fusion is another similarity within the analyzed application. Very different methods and techniques for fusing information are applied: methods from classical mathematical approaches like Dempster-Shafer and Bayes to knowledge-based techniques like structural models [4] are applied for information fusion.

The identified monitoring tasks, derived from the set of analyzed monitoring systems are shown in figure 6.

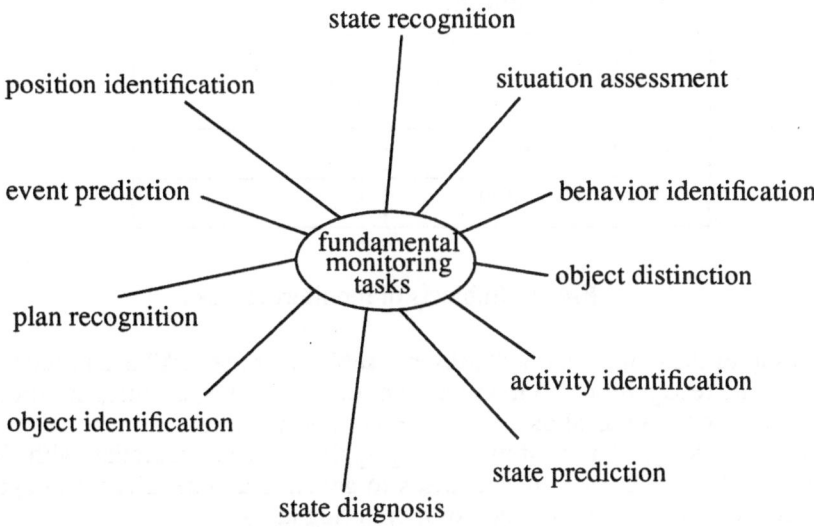

Fig. 6: Fundamental monitoring tasks

How can we use the knowledge about the fundamental tasks of monitoring applications to create a generalized monitoring model? The generation of the Category Model is one answer to that question.

3. The Category Model

Looking at the set of fundamental monitoring tasks, one may mention that between these tasks no relation do exist. The relation between the tasks and the subjects of the tasks help to understand the step towards the Category Model. For example, the subject of the task "state recognition" is the "state" of an object or a system. The subject represents "what" has to be monitored. The relations between these tasks and their subjects is given in the following figure 7.

monitoring task	subject of the task
state recognition	state
situation assessment	situation
behavior identification	activity
activity identification	activity
state diagnosis	state
plan recognition	activity
position identification	state
state prediction	state
object identification	object
event prediction	event
object distinction	object

Fig. 7: Subjects of monitoring tasks

Result of the confrontation "tasks <-> subjects of the task" is a limited set of subjects. The recognition and monitoring of objects, states, events, activities and/ or situations is a main goal of most of the monitoring tasks.

With the knowledge, that most monitoring applications are operating with similar subjects of their tasks, the chance arises to generate a generalized homogenous model for representing the subjects of monitoring tasks.

In many cases a model for representing and handling the subjects of monitoring tasks is called the *world model* [13 (page 908), 14 (page 10)] of the monitoring system and serves as an representational framework.

These subjects are representing the categories, the basic elements of the Category Model on which the monitoring applications operate.

The Category Model consists out of the categories identified in an analysis of monitoring applications and the relations between the categories. A top level view of the relations is represented in figure 8.

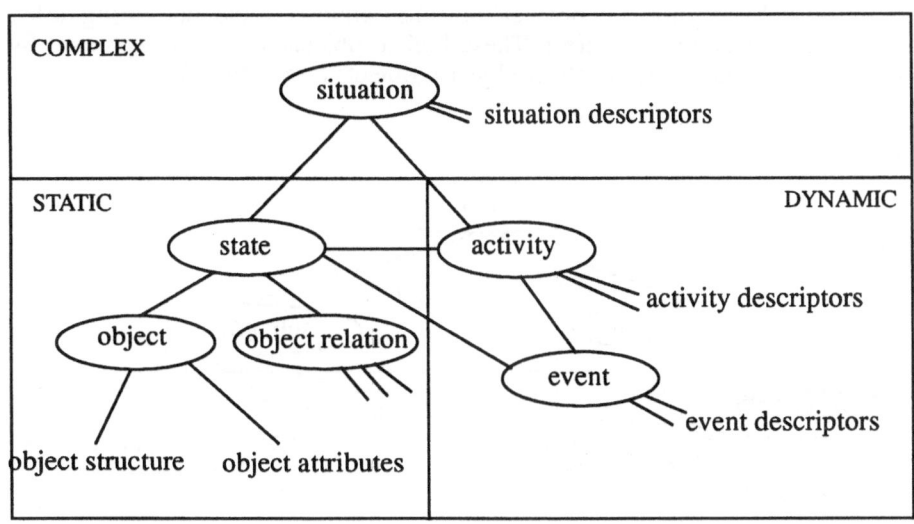

Fig. 8: The Category Model

Underlying the Category Model, three aspects characterize the categories of the model: the *static aspect*, the *dynamic aspect* and the *complex aspect*. By these aspects, the categories become structured in their temporal meaning. While the objects can be understood as static parts of the monitoring tasks, the dynamics are represented by events or activities, creating the dynamic part of the model. The complex aspect, represented by the category situation, combines the polarity of the static and dynamic aspect. Elements of the category situation are subjects of monitoring tasks, where the goal is to identify a situation composed out of states and activities. Examples for each category will be given below.

In this context, the meaning of static is that the subject of the monitoring tasks, i.e. the elements of the object category, is more or less invariant in time, while elements of categories like events or activities may change in time.

The Static Aspect

The static aspect operates on the physical and logical states and objects of the application area. While in the air traffic management domain the aeroplanes are the objects, in the indication & warning application domain i.e. persons are the objects to be monitored. Extending the goals of the monitoring tasks of some domains, not only the objects in a whole, but also the identification of structures and attributes of objects is of relevance.

Object structures describe defined structures of an object, i.e. a physical layout of entities in the environment. *Object attributes* describe selected attributes of the object like the color or the size. In some applications, i.e. tactical situation assessment systems, the object relations between objects are of interest. Herein it is important to handle not only physical object as subjects of the monitoring tasks; also logical objects are of interest. These logical objects are often expressed by hierarchical classifications of sets of physical objects (see figure 9).

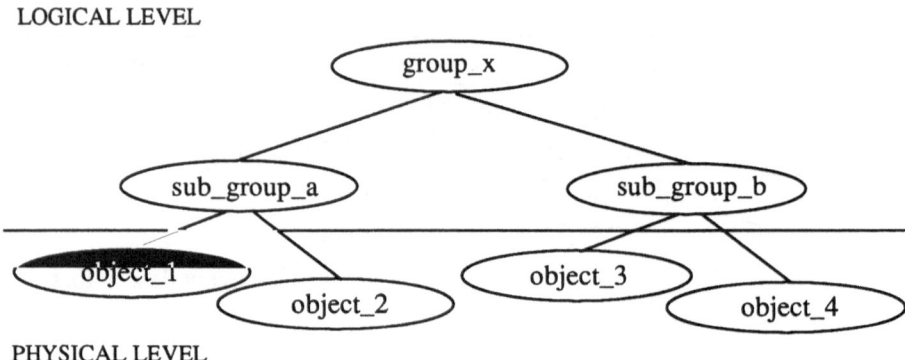

Fig. 9: Logical objects composed out of physical objects

While the objects of the physical level can be detected by sensors, the objects of the logical level can be detected using the following two methods: The first method is to infer from detected physical objects to the logical objects. This process can be activated, if the evidence for the existence of the physical objects is high enough. The other method is to extract the information directly out of the input stream leading into the monitoring application. In this case the input message comes from i.e. a human source and contains the information about the logical object in a symbolic encapsulation.

Both categories, objects and object relation, build the category state, which represents the set of all static aspects of the environment.

The Dynamic Aspect

The dynamic aspect of the Category Model includes everything related to activities, plans or other time-consuming events. Goal of the monitoring of dynamic aspects is the recognition of specific events, plans or activities. The relation between events and activities is time-oriented and is based on the assumption, that the duration of events is much more smaller than the duration of activities. Events are of very short duration and represent a basic activity unit, while activities are composed of one or more events (see figure 10).

The absolute time scale for the definition what an event or an activity is, depends on the time scale the application requires. In an automation application the duration of events may range from milliseconds to minutes, while in tactical situation assessment systems events may range up to hours.

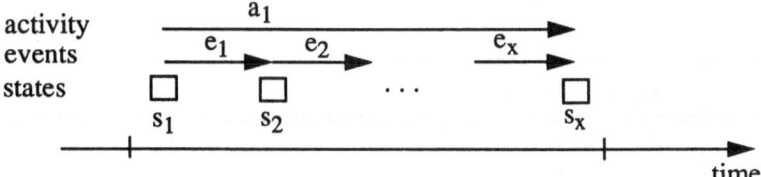

Fig. 10: Relation between static and dynamic aspects

The Complex Aspect

The category situation represents dynamic aspects as well as static aspects. In many applications, the identification of a situation is required. This means that not only recognition of an object or an activity is requested. To identify a situation means to recognize a configuration composed out of objects and activities. For this reason, the category situation has been introduced to handle those cases, where the subject of the monitoring task is described by a combination of states and activities.

An Example

An example taken out of the air traffic management domain describes what the goal and subjects of monitoring tasks are and what an approach for a representation within the Category Model is. Two typical tasks of an air traffic management subsystem are:

1. Monitoring of the separation (state) *between two aeroplanes* (objects).
2. Recognition (of the situation) *that two aeroplanes* (objects)

   ... are on collision course (activity).

   ... come into collision in a defined time or place (event)

The mapping of the subjects of the monitoring tasks onto the categories of the Category Model is given within the brackets.

The first task describes the monitoring of the distance between two aeroplanes. The distance, which is not allowed to become smaller than a defined minimum value, represents a state, depending on the spatial object attributes.

The second tasks represents the problem in two variations. While the first variation is described as an ongoing activity, the second variation describes the same problem as an event-oriented case. In each variation, the recognition of the collision situation is composed out of a static category element - the objects - and a dynamic category element - the activity respectively the event.

These examples show, how monitoring requirements can be interpreted in terms of the Category Model. The set of available categories allows to capture a very broad range of monitoring tasks. An example for representing the elements of the categories is given below:

category	representation
object	aeroplane(F16,...) aeroplane(B747,...)
state	*...objects with spatial, temporal attributes* aeroplane_state(aeroplane( ),position(), time, ...)
event	*... as basic activity units* event(x,...)
activity	act_takeoff(...) act_flight(event(start_a), event(dest_b), path, aeroplane(..) act_approach(...)
situation	sit_collision(act_flight(...B747...), act_flight(...F16...))

Fig. 11: Representation of category elements

4. Conclusions

Based on the requirements and tasks of different monitoring applications, fundamental monitoring tasks have been extracted. The similarities between the subjects of the monitoring tasks lead to a set of subject types, called categories. These categories have been refined and structured into static, dynamic and complex aspects which build the Category Model.

With the Category Model, a framework for the representation of the subjects of monitoring tasks is given. The process of information fusion uses the model for classifying input in finding a category for different types of information. While numeric data can be used for enhancing the attributes of the category elements, symbolic information, as provided by human sources, can directly represented as an category element.

The Category Model, as described in this paper, is the structural foundation for analyzing monitoring tasks and representational stage of the information fusion process.

The Category Model is a part of a generalized analysis model for monitoring tasks and one step towards a generic monitoring architecture using knowledge-based techniques.

Acknowledgments

Several results of this work have been developed during the membership of the author in the ESPRIT II project 5345 DIMUS - Data Integration in Multisensor Systems at the Institute for Computer Science of the Technical University of Munich. For the discussions during that time, the author would like to express his appreciation to Silvia Pfleger and Bernd Radig.

References

1. Hansen, C., Henderson, T.C., Shilcrat, E., "Logical sensor specification", Proceedings of the 3rd International Conference of Robot Vision and Sensor Systems, pp.321-326, 1983
2. Wechsler, H., "Computational vision", Computer Science and Scientific Computing, pp.406-421, Academic Press, 1990
3. Miles, J.A.H., Faulkner, H.C., "Knowledge-based techniques for tactical situation assessment", Conference Proceedings MILCOMP 88, pp.313-318, 1988
4. Waltz, E.L., Buede, D.M., "Data fusion and decision support for command and control", IEEE Transactions on SMC-16, no.6, pp.865-879, Nov./Dec. 1986
5. Wilson, G.B., "Some aspects on data fusion" in Advances in Command, Control & Communication Systems, editor Harris C.J., pp.321-338, P.Peregrinus, London, 1987
6. Azarewicz, J., Fala, G., Heithecker C., "Template-based multi-agent plan recognition for tactical situation assessment", Proceedings of the 5th conference on Artificial Intelligence applications, pp. 248-254, Miami, March 1989
7. EUROCONTROL, "MADAP-System Descripition", Issue 02, EUROCONTROL Brussels, Belgium, 1986
8. BFS, "Flight Track Monitoring System FTMS - Functional System Description", Bundesanstalt fuer Flugsicherung, Frankfurt a.M., 1984
9. Jakob, F., Suslenshi, P., "Situation assessment for process control", IEEE Expert, vol.5, no.2, pp.49-59, 1990
10. Luo, R.C., Lin, M.-N., "Hierarchical robot multi-sensor data fusion system", NATO ASI series, vol.58, pp.67-86, 1988
11. Lenat, D.B., Clarkson, A., Kiremidjian, G., "An expert system for indication & warning analysis", IJCAI-83, pp.259-262, 1983
12. ESPRIT Project 5345 "DIMUS - Data integration in multi-sensor systems", "1st Design Specification Report",DIMUS/12/dv/0001-01/b/P1, 12.09.91
13. Luo, R.C., Kay, M.G., "Multisensor integration and fusion in intelligent systems", IEEE Transactions on SMC-19, pp.901-931, 1989
14. Flynn, A.M. "Combining sonar and infrared sensors for mobile robot navigation", IEEE International Journal of Robotics Research, vol.7, no. 6, pp.5-14, 1988

A flexible Real-Time System for Vessel Traffic Control and Monitoring

Livio Stefanelli
Rigel Engineering S.A., Av. Rogier 385, Brussels
Belgium

Abstract

Surveillance systems are increasing their complexity in terms of real-time requirements, performance, sensor types and multiplicity, human computer interface and dependability. To this category belong Vessel Traffic Control Systems (VTS) that support control and monitoring of ship traffic in congested areas. The ESPRIT project 6373 TRACS, partially funded by CEC, is developing a new generation of VTS based on the exploitation of high performance computer and intelligent data processing & monitoring techniques. Keywords: surveillance system, VTS, distributed system, heterogeneous sensors, image processing, tracking, parallel architecture, data fusion, human computer interface, scene presentation, real-time system.

1. Introduction

The new class of VTS systems, that will be carried out under the TRACS project, is dedicated to the monitoring of complex vessel traffic in well defined areas with a high density of dynamic (e.g. cargo, tug) and static (e.g. buoys, platforms) objects. A VTS has real-time constraints less strict than an Air Traffic Control system, nevertheless the objects under control, in general, have to be considered as passive objects because no electronic code (target identifier) is transmitted automatically by the target. For this reason, a sophisticated ground based radar is generally used. In some case, it can be necessary to complete the information captured by radar with additional sensors such as video camera, sonar, etc. In general, VTS systems can be classified according to adopted sensor(s) and the complexity of the real-time processing system housed in the port control centre:
- VTS installations with one or more sensors that can belong to homogeneous or heterogeneous types (e.g. radar, radio direction finder, low light camera, GPS).
- Centralised or distributed architecture: two basic possible choices for the VTS

processing system that can be based on conventional or high performance computers (i.e. RISC, parallel machine).

- Software complexity mainly depends on the implementation of data processing, data fusion and pattern recognition techniques, image processing, real-time features (e.g. hard real-time constraints), human computer interface and knowledge based subsystem to support navigational assistance service.

TRACS implementation exploits advanced solutions such as the support of homogeneous or heterogeneous radar sensors, distributed system architecture based on high performance computer, data fusion and new pattern recognition techniques, sophisticated man machine interface and hard real-time features.

2. System Overview

The TRACS prototype is derived from a more general system architecture that can support a wide range of advanced surveillance systems. The "TRACS generic system architecture" has been defined to satisfy an architecture open and sufficiently general to support a subset of different surveillance systems such as VTS, ATC systems, environmental monitoring systems. Each component of this generic architecture carries out a set of specialised tasks, the major building blocks of TRACS generic architecture are basically six (see figure 1):

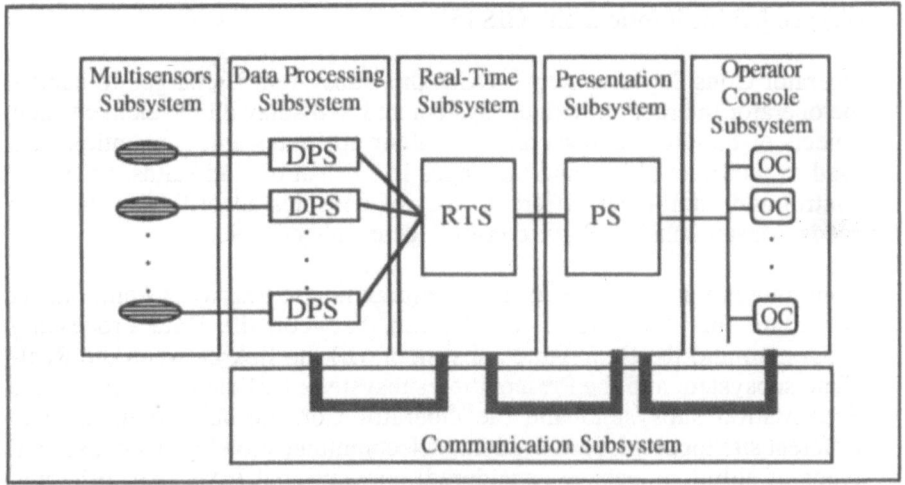

Fig. 1: System overview

i. Multisensor Subsystem: this component, based on several and different types of independent and heterogeneous sensors, is able to capture, under different atmospheric conditions, the target data (e.g. radar equipment).

ii. Data Processing Subsystem - DPS: each sensor output is processed by a specific data processing component based on a parallel architecture. This subsystem is scalable and extensible in accordance with the expected maximum number of objects under control. Different types of sensors could be controlled by different types of data processing components (data processing heterogeneity) able to support different operational requirements. The different data processing units implement specific data & image interpretation algorithms to transform the sensor raw data into synthetic data. In fact object features are extracted applying bidimensional and/or 3D image analysis techniques.

iii. Real Time Subsystem - RTS: this subsystem is responsible if all the real-time processing tasks and is based on a dedicated real-time machine. The synthetic data (.i.e. DPS plot) are processed according to specific 2D/3D data fusion techniques to determine the shape, speed and to check specific navigation constraints. In case of constraint violation different alarms are generated. A real-time operating system provides specific services and mechanisms to suport the RTS applications, typically: advanced IPC (e.g. broadcast messages, RPC, port migration), tasking facilities (e.g. threads) and hard real-time features (e.g. deadline handling for periodic and sporadic task).

iv. Presentation Subsystem - PS: the Presentation Subsystem is charged to handle all the real-time data elaborated by RTS. Three basic data types can be identified: (i) Messages, alarms, reports; (ii) static and dynamic object; (iii) geographic information - GIS [5].

v. Operator Console Subsystem - OCS: this subsystem is charged to handle the operator interaction with the system and to display all the elements and objects of the areas under control. Colour attributes and conventions are used to facilitate object and messages identification and status. OCS can control and administer different types of consoles according to the user needs: Master console, Slave console, Read-only console.

vi. Communication Subsystem: three communication/network components constitute this subsystem: (i) The link between the Data Processing subsystem and the Real-Time subsystem, (ii) the link between the Real-Time subsystem and the Presentation subsystem; (iii) the link between the Presentation subsystem and the Operator Console subsystem. Due to different site topologies and National telecommunication legislation, several types of solutions can be considered: conventional LAN (e.g. Ethernet, Token ring), X25 private line, Hertz communication on available transmission bands, fibre optic approach, etc.

3. Data Fusion and Scene Comprehension

The multisensor tracking (i.e. multiradar tracking), the data fusion and scene comprehension core modules allow the logical integration of different sensors and a correct target interpretation into a VTS system. A centralised or a distributed approach of Data Processing Subsystem can introduce important repercussions on the implementation of multiradar tracking and data fusion components [1].

The centralised solution consists of the direct connection between sensors and a centralised DPS charged to implement the multiradar tracking (see part left figure 2). In this case, only an homogeneous set of sensors could link the DPS component. The resulting computational load can be easily supported by a conventional DPS machine.

In the distributed solution, adopted in TRACS, each sensor is supported by a dedicated DPS that is linked to the real-time subsystem (i.e. RTS) charged of data fusion computation and scene understanding (see figure 2).

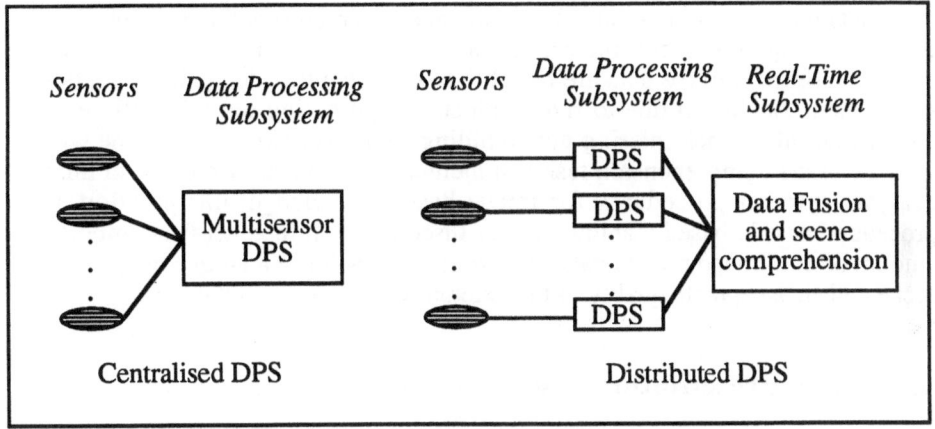

Fig. 2: Centralized and distributed data processing subsystem

Major benefits of this approach are: integration of heterogeneous sensors, availability of an increased processing power to support 2D (or 3D) image analysis, simplicity and system scalability, transmission of synthetic data instead of sensor plot and row data. Each DPS extracts object shape, size, position, direction, yaw, barycentre and tracking information from the row data of the sensor. It is worth to notice that a conventional DPS exploits row data only to compute the target barycentre with a considerable loss of sensor information.

These 2D monotracking information are elaborated exploiting classical image processing algorithms [2] such as: filtering (i.e. sea and atmospheric clutter), adaptive thresholding for image digitalisation, morphological operations, edge extraction, edge following, medial axis transformations, thinning, segmentation, region growing.

The processing power of a DPS is essentially related to the maximum number of expected targets and to the sensor scan time [2], for this reason, the control of areas with congested traffic requires DPSs equipped with parallel and scalable machines. As the monosensor tracking facilities are provided by DPSs, the final track combination, data fusion computation and constraint check are performed by RTS. The data fusion processing produces two types of information on the identified targets: dynamic data such as the speed of the ship and static data such as final barycentre position expressed in Cartesian coordinates, direction, yaw, speed shape length and shape co-ordinates. Dynamic and static data are independent from a specific sensor and from a particular detection or scan time (logical scan time vs. physical scan time) [3], [4]. The accuracy of the fused data is better than the traditional multiradar tracking (figure B) and RTS can also support, extrapolation of the target estimated position and track (i.e. dead reckoning) in case of missing or insufficient data.

Another basic RTS functionality is to check the fused data against static and dynamic traffic constraints in order to send warnings and alarms to the operator consoles. Typical geographic constraints are security distance from platforms, buoys, private or military areas, line of depth, anchored ships, obstacles, etc.

Data fusion techniques and constraint analysis represent the basic blocks to achieve a complete scene interpretation, especially in circumstances where processed or incomplete data are not able to solve context-dependent situations such as ship occlusion due to fixed objects, cargo hiding tug or small vessel, track restart after track missing due to hiding, ship overtaking, etc. In the worst case the scene comprehension task can include hierarchical or multilevel image analysis, general procedures for image-based knowledge manipulation and procedural or rule-based manipulation of fused data [3]. In general, an important amount of non-image related data are processed to achieve a unique image of the scene and an acceptable quality of the external world representation.[3].

4. Scene Presentation and Rendering

The usability of a VTS depends on the quality of the scene presentation, object rendering and operator interaction style. For this reason TRACS console provides several functions such as zoom, windowing, scaling, range marks, antenna sweeps, acoustic and colour attributes, the bow and the stern attributes, form fill-in, menu bar, pull-down menu, buttons, dialogue box and selection fields. In TRACS, the operator demands information on displayed objects by using the "object-action" style (i.e. Macintosh style).

This approach presents three major advantages[4]: (i) small memorisation of actions and no error in selecting operations, (ii) limited choices of permitted operations. All commands applied to selected objects are in the menus and forbidden actions are removed or not accessible, (iii) the user has the complete control of the applications after a short training. Each presented object (e.g. ship) has generally three types of attributes: static (identifier, dimension, type, flag, etc.), dynamic (e.g. position, speed, direction angle, yaw angle, shape), operational (e.g. voyage number, hold condition, expected mooring place, carried goods, etc.) [4].

Additional information can be requested to external databases or repositories. In this case the data are displayed on a dedicated area of the screen. In fact the screen dedicated to the scene presentation must be visible to the operator all the time and window overlap is forbidden.

5. Conclusion

The goal of the final TRACS prototype is to carry out a new generation of VTS to overcome the typical limitations of current systems that consider the target as a single point and the form and the shape are not displayed with an important loss of physical and logical information. In fact, a complete information on the targets can partially support an automatic control of the ship behaviour and some critical functions such as control of departure and arrival operations, control of ship paths and speed limits, cross distance estimate and automatic detection of collision avoidance, stranding avoidance, intrusion avoidance, etc. The representation of the scene elaborated by the TRACS distributed system will be displayed on a colour high resolution screen.

A sophisticated man-machine interface will allow to a small number of operators to monitor complex traffic scenario, to recover dangerous events and to demand ship information to external databases or repositories. Following the "TRACS generic architecture" definition, the hardware and software components of TRACS prototype are:

- *DPS*: INMOS Transputers (T800 or T9000),
- *RTS*: RISC machine (i.e. SUN Sparcstation) and CHORUS OS,
- *Presentation & Operator Console subsystems*: RISC machine (i.e. SUN Sparcstation) and UNIX. As far as concerns the Human Machine Interface, the graphical and windowing system is based on X Window defined by the X Consortium with the 'Look and Feel' of OSF/Motif.

The project, started in June 1992, will be completed in November 1994, a first prototype will be available during the first quarter of 1994.

References

1. A. Farina, S. Pardini, "Introduction to Multiradar Tracking Systems", Rivista Tecnica Selenia, Vol.8, No. 1, pp. 1-13, 1991.
2. S. Bottalico, F. De Stefani, M. Spada, "Image Processor Design Report", ALENIA, TRACS deliverable WP5, Nov. 1992.
3. S. Bottalico, F. De Stefani, M. La Manna, A. Righetti, "Application Specification and User Recommendations", ALENIA, TRACS deliverable WP1, Aug. 1992.
4. Y. Urbain, "Human Machine Interface Specification", Rigel Engineering, TRACS deliverable WP4, Nov. 1992.
5. L. Stroobants, "State of the Art of Geographic Information Systems and System Specifications for Integrability", Rigel Engineering, TRACS deliverable WP4, Nov. 1992.

ESPRIT Project AZZURRO:
Data Fusion for Marine Protection

Susanna Ghelfo
Agusta Sistemi
Tradate (Varese) Italy

Abstract

This paper will present the objectives of the AZZURRO project from the application point of view. This project is devoted to the development of advanced software techniques for the fusion of environmental data collected by two different sensors: an active sensor such as a Lidar Time Resolved Laser Fluorosensor (TRLF) and a passive one such as a multispectral scanner (Daedalus).
The main interesting aspects of this work will be:
- to prove if adequate data fusion techniques are able to implement the synergy existing between the measurements of the active and passive sensors;
- to develop data presentation methods acceptable by the final users;
- to demonstrate that the resulting thematic maps increase the knowledge of the marine environment providing also input for intervent actions.

AZZURRO is an Esprit III project (P7207). Agusta Sistemi(I), Onera(F), KUL(Katholieke Universiteit Leuven)(B), Piaggio(I) and Steria(F) are the partners involved in it. It will last three years starting from the 1st of October 1992.

1. Introduction

In the last 20 years, the observation of the marine environment, previously performed by research vessels through punctual measurements [2], has evolved versus a synoptic observation made by air-space borne passive sensors.

The application of remote sensing, whether from satellite or airborne, to monitor the water-quality offers significant advantages over direct sampling and local monitoring techniques. In a period of a few minutes, or at most a few hours, the equivalent of tens or even hundred of thousands of sample measurements can be taken, covering regions many hundreds of kilometers in extent. The cost and logistics of gathering and analyzing such a large number of samples by conventional sampling methods is not a realistic proposition. Consequently surveys attempting to characterize large water bodies frequently rely on bottle samples and direct measurements in the water column, sometimes gathered days or even weeks apart, from widely dispersed locations. Attempts to predict the nature, dynamics and biomass productivity of large water bodies based on such irregular data are, at best, open to significant error.

Clearly passive remotely sensed data such as multispectral imagery produced by either airborne or satellite sensors, which measure the backscattered solar irradiance from the surface water, can only provide information about this upper layer. This limitation can be overcome with a few ground-truth measurements made at strategically located sites within the survey region.

A Time Resolved Laser Fluorosensor (TRLF) helicopter-borne, constitutes a system that can perform "sea-truth" campaigns, in place of vessels, measuring Phytoplankton, Total Suspended Matter and Dissolved Organic Matter and giving their profile in the water column. The complementary nature of simultaneous measurements (by Lidar and Daedalus) of independent but related parameters enables many of the limitations of each single instrument to be reduced and to eliminate many of the ambiguities encountered in the interpretation of any one data set.

Having in mind these considerations, the project is oriented to:
- overfly a marine area at the same time with an airborne passive sensor, Daedalus multispectral scanner, and an helicopterborne active one, LIDAR-TRLF;
- develop data processing methods able to assign a correct interpretation to the single measurements;
- fuse Daedalus and Lidar data in order to produce a thematic map of the overflown area.
- validate the remote sensed measurements by sea-truth campaigns

The final output of the project will be a software system prototype able to process and fuse the data collected by the Lidar-TRLF and the multispectral scanner Daedalus and to provide thematic maps that give a useful knowledge of the marine environment.

2. Functional Description of the System

An helicopter equipped with a Lidar-TRLF and an airplane equipped with a multispectral scanner Daedalus (see Fig.1) will overfly and sense, at the same time, some marine areas in the Northern Adriatic basin. Moreover, sea-truth measurements will be performed in a few specific points inside the overflown area. These measurements will be used to validate the data obtained through remote sensing operations and data fusion. The input data for the system will be:
- Lidar and Daedalus data
- local measurement campaigns data

The sensor data will be available, on magnetic tapes, for processing a few hours after the gathering time. The pre-processing Lidar and Daedalus modules will derive a set of data starting from the registered signals. They will apply on the data the necessary geometric and atmospheric corrections and will georeference the single measurements.

The data validation and verification module will check the consistency between the Lidar and Daedalus data. The data fusion module represents the core of the project. A set of data fusion methods will be developed or adapted to our application so that to implement the existing synergy between the two sets of sensor measurements.

The controller will coordinate the interaction between all the modules of the system.

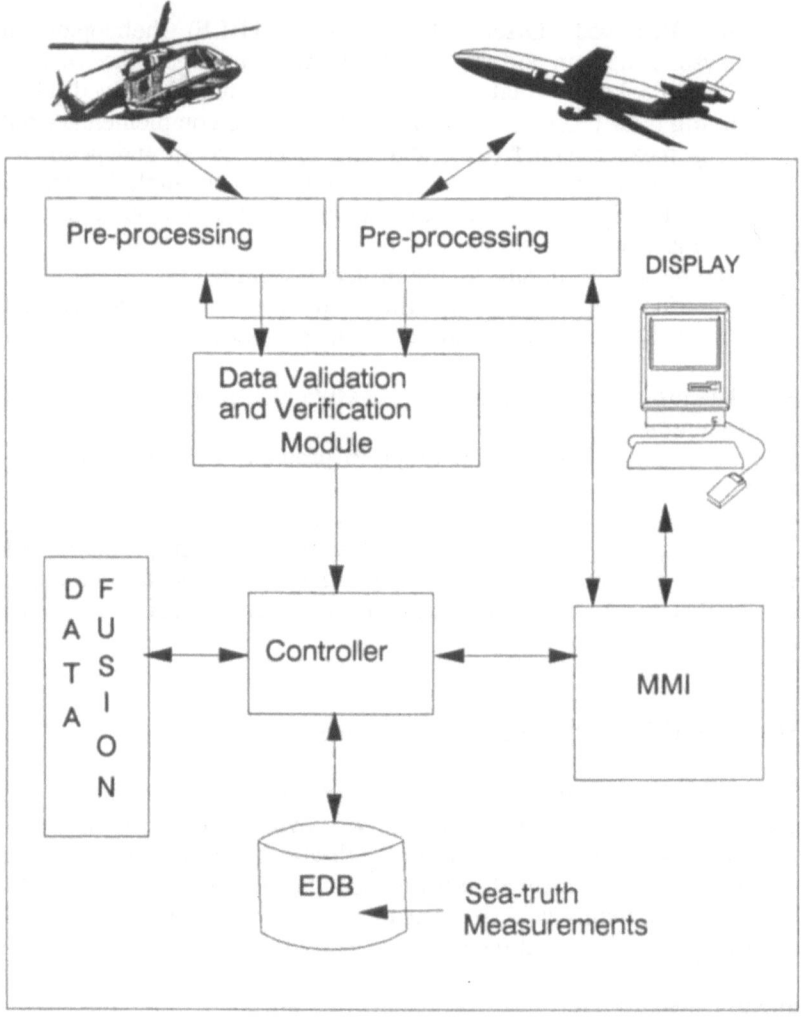

Fig. 1: Data flow of AZZURRO system

The Man-Machine-Interface module (MMI) will manage the interaction between the system users. It is possible to distinguish 3 types of users, one for each step of data processing:
- pre-processing user
- fusion user
- final user (oceanographers, authority devoted to the marine protection, etc.)

For each type of users an ad-hoc interface will be developed. The External Data Base module (EDB) is the data base containing information such as:
- sea-truth measurements
- geographic information, etc

3. Performance of the Lidar and Daedalus Sensors

The multispectral scanner Daedalus performs its measurements using the upwelling solar radiation from the sea. After a careful subtraction, from the content of each spectroscopic channel, of the solar radiation scattered by the atmosphere in the sensor field of view, it is possible to determine the various substances present in the water using the content of some channels.

The empirical procedure to derive the retrieval algorithms of such substances [1] is based on the use of the sea-truth measurements. From the Daedalus data, the measurements of the following water quality parameters will be derived for this project:
- the Total Suspended Matter (TSM) concentration
- the Dissolved Organic Matter (DOM) concentration
- the Chlorophyll-a concentration
- the water turbidity index (diffuse attenuation coefficient)

The Lidar-TRLF performs its measurements using the results of the interaction mechanisms of a laser light with the water and its constituents. These mechanisms are mainly of two types:
- absorption
- scattering

The absorption generates fluorescence phenomena correlated to the presence of DOM and chlorophyll-a in the water. The scattering phenomena are due essentially to the presence in the water of TSM and pigments. From the spectro-temporal analysis of the water response functions it is possible to derive the substance concentrations and their profile in the water. The measurable water parameters used in this project are the same as Daedalus but they are derived using completely different light-water interaction mechanisms.

Moreover, the Lidar-TRLF provides the profile of the substance along the water column until a depth which depends on the water transparency and on the laser emission wavelength.

4. Lidar-Daedalus Synergy

Lidar and Daedalus measure respectively inherent and apparent optical properties of the water. According to Preisendorfer-65, the optical properties of a medium can be classified according to their invariance with respect to changes in the radiance distribution. One distinguishes "inherent" properties which are independent from radiance distribution and "apparent" properties that change with radiation distribution.

Using this two optical characterizations of the water body it is possible to develop a methodology for using the data of both sensors in the derivation of the final thematic maps. In fact, the retrieval algorithms for the Daedalus data processing depend on the characteristics of the water body. As much as these characteristics are precise, more correct are the information derived by the Daedalus data. The precision and quantity of the Lidar data allows us to adapt the retrieval algorithms according to the variations of the water body characteristics inside the entire survey region instead of a few located sites.

This is a new approach in the Daedalus data interpretation that allows us to reduce the uncertainties associated to the final thematic maps.
All the data fusion methods that will be developed should base on this synergic use of the two data sets.

5.   Thematic Maps and Man-Machine-Interaction

A set of sea thematic maps will be generated by the developed software system. A sea thematic map, with respect to a parameter p, is a two(three)-dimensional representation of the sea, generally built on the cartographic description of the area and consisting of various colorful zones. Different colorful zones represent areas with different values for the parameter p.
The following thematic maps will be derived:
• TSM concentration
• DOM concentration
• Chlorophyll-a concentration
• Water Turbidity
• Water body Homogeneity
It is very important, for the final user, to have:
• the final thematic maps in "quasi-real time" respect to the remote sensing activity
• an easy-comprehensible thematic map presentation (2D and 3D maps with zoom facilities, etc.)
• a user-friendly graphic interface with the system (menu and/or icon based)
• browsing facilities to extract all the information associated to a map (inspect the map, point by point showing the vertical profile of the substance in water, the homogeneity of the associated water column, etc.).
For these reasons, a big effort will be made by the Partners on the design and development of an advance software package for the final data presentation.

6.   Validation of the Results

The software system will be tested on the data derived from an in-field campaign on some selected marine zones. These zones must be characterized by strong gradients in the concentration values of the marine parameters defined above. Such zones can be found around the Po river delta and along the Emilia Romagna coasts. The selected areas will be sensed "contemporarily" by the Lidar-TRLF on the Agusta helicopter AB412 and by the Daedalus multispectral scanner on the Piaggio airplane P166. At the same time, sea-truth measurements will be performed by research vessels.

7. Conclusions

This project has been considered of great interest by a few possible final users (authorities devoted to the marine protection, etc.) who appreciated the effort made to integrate the data. In fact, a lot of water monitoring instruments exist, most of them giving partial information. Of course, this situation is rather difficult for the final user that is overwhelmed by thousands of data not easy to understand and to use.

The AZZURRO project could represent a promising step both for the final users and for the information technology scientists that could be interested in the pure data fusion techniques.

References

1. Sturm-92: B. Sturm "Ocean Color Remote Sensing: A Status Report" (private comunication)
2. R.W. Preisendorfer "Radiative Transfer on Discrete Spaces", Pergamon Press, New York (1965).

ESPRIT Project DIMUS:
Data Integration in Multisensor Systems

F. Benvenuto, M. Ferrettino, M. Pasquali, F. Perotti, P. Verrecchia
Ansaldo Ricerche S.r.l.
Corso Perrone 118, 16161 Genoa, Italy

Abstract

The surveillance problem in the transport field seems to be quite critical due to the difficulty of controlling simultaneously different situations and environments.

Ansaldo Ricerche is participating to the realisation of a system capable of supporting the activity of a human operator in the surveillance of underground station environment. The first results of the project, called DIMUS, are shown in a demonstrator system that has been realised in a laboratory environment.

The demonstrator is able to perform:
- the environment monitoring with different sensors (TV cameras, microphones, photocells, tactile arrays, etc.);
- the processing and the fusion of the data coming from the sensors in order to detect the dangerous situations;
- the management of the interaction with the operator.

The project is carried out by a consortium of European Companies, Universities and Research Centres with Ansaldo Ricerche as main contractor and is partially founded by CEC ESPRIT programme (Project 5345 and 7809).

The paper shows the functionalities and the architecture of the DIMUS demonstrator, underlining the technologies used.

1. Introduction

The underground transport systems trend is to have automatic and unmanned plants without neglecting passengers' surveillance and safety. At present, TVCC systems are used in order to control station platforms; other devices used for passengers safety are emergency panels situated on the platforms.

The visual information are sent to the operator central room, where an array, often large, of monitors allows to show them, in a cyclical way. The operator's task is to detect potentially dangerous situations and to react appropriately. The major problem in such vigilance tasks is that the human operator is incapable of maintaining optimum performance over some arbitrary duration. One reason for this is that in certain situations the operator activity to process incoming information is overloaded, often resulting in the loss of crucial information and a decrement in performance. A second reason is that in other situations the operator is under loaded cognitively and can become bored or fatigued, allowing himself to be easily distracted from the task in hand. The DIMUS system is intended to counter both of these problems.

2. The DIMUS project

Ansaldo Ricerche is participating to a European project called DIMUS (Data Integration in MUltisensor Systems, P5345 and P7809), partially funded by CEC within the ESPRIT program (whole value of the project 7000 KECU), for the development of an information processing system capable of supporting the activity of a human operator in the surveillance of underground station environment. The DIMUS [1] system reduces the overall information processing load on the operator, and at the same time, taking over the usual aspect of the vigilance task, directs the operators attention to only the information most relevant to a potentially dangerous situation. To perform such work the DIMUS system is provided with the following capabilities:
• monitoring of the environment with different set of sensors (TV cameras, microphones, photocells, tactile arrays, etc.);
• processing and fusion of the data coming from the sensors in order to detect the dangerous situations;
• management of the interaction with the operator.
    The DIMUS project [2, 3, 4] is developed by a consortium formed by the following partners:
• Ansaldo Ricerche s.r.l., Genoa (I), as Main Contractor
• IRST - ITC, Trento (I)
• University of Genoa (I), Depart. of Biophysical and Electronic Enginering
• Elsag Bailey, Genoa (I)
• Signum Computer GmbH, Munich (D)
• Thomson- CSF LER, Rennes (F)
• TUM - Technischen Universitat Munchen, Munich (D)
    The DIMUS project started on February 1991; the first phase is finished on October 1992 with the realisation of a demonstrator system operating in a laboratory environment. A second phase of the project is foreseen and its goal is the installation of the demonstrator system in a station of Genoa Underground.

3. The DIMUS demonstrator system

The DIMUS demonstrator system has been realised in the laboratories of Ansaldo Ricerche in Genoa, where two environments have been created in order to simulate both an underground platform (see figure 1) and an underground central control room. The software has been developed on a SUN

Sparcstation 2, with SUN/OS 4.1.1 operating system, using C and C++ language (together with SUIT library for the graphical user interface).

The laboratory demonstrator system is able to execute the following surveillance tasks:

- detection of people in prohibited areas (as tunnel, etc.);
- detection and localisation of anomalous acoustic events (as scream, gun shot, etc.);
- detection of people in dangerous zones (as platform border, tracks, etc.);
- evaluation of level of crowding, with detection of overcrowding situations.

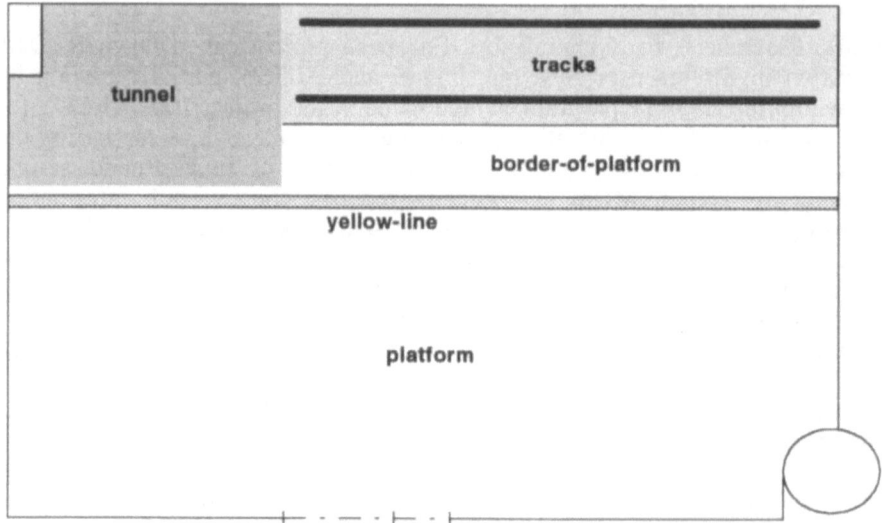

Fig. 1: DIMUS demonstrator system, simulated underground platform

The detection of a dangerous situation is pointed out to the operator through a graphical user interface on which both the kind of alarm and the position of the anomalous event is shown. The attention of the operator is then focused on a monitor showing the detailed image of the anomalous event coming from a mobile camera. Additionally the operator has the capability to drive the mobile camera with a joystick in order to improve his understanding of a scene.

The hardware architecture of the DIMUS demonstrator system is shown in figure 2 and is composed by the following elements:

- a SUN Sparcstation 2;
- a VME-Bus based Acquisition Subsystem connected to the Sparcstation through a special bus adapter board;
- two VDOT32 Digital input/output boards;
- a Series 150 image acquisition subsystem from Imaging Technologies;
- a video multiplexer 8 to 1, controlled by the software modules through the VDOT32 boards;
- three fixed colour cameras and one mobile b/w camera connected to the Series 150 image acquisition subsystem and to the video multiplexer;

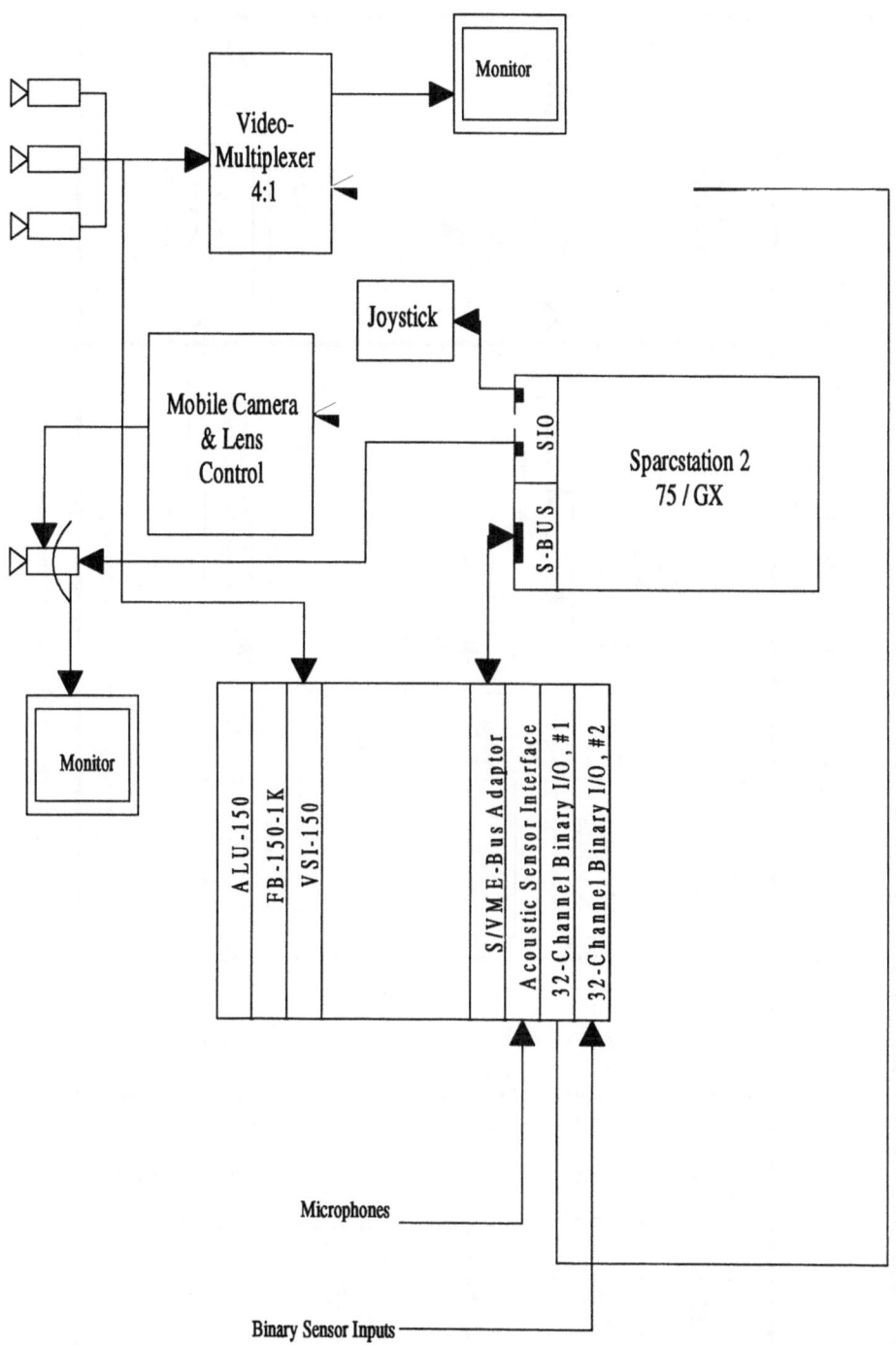

Fig. 2: Architecture of DIMUS demonstrator system

54

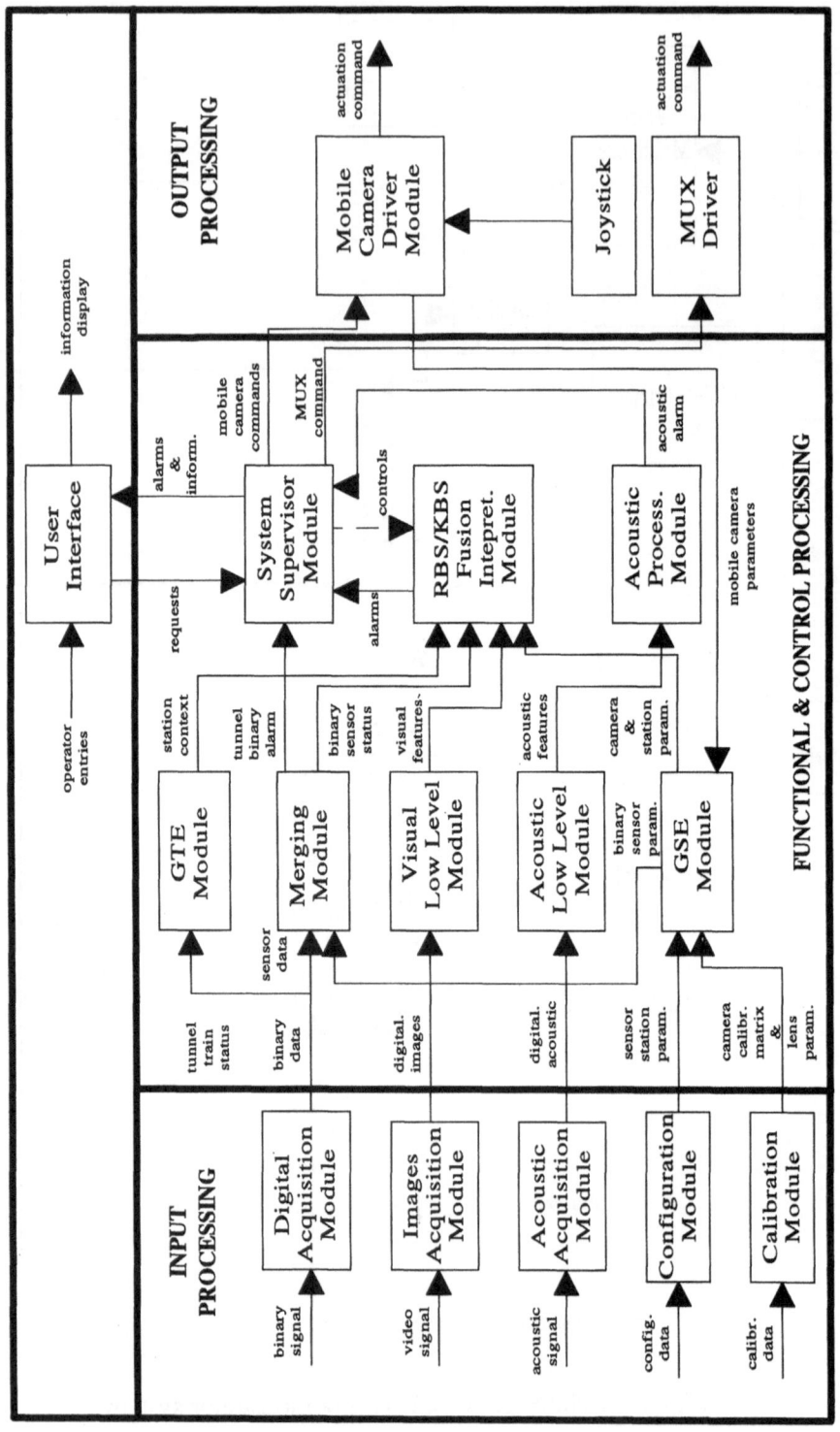

Fig. 3: Software structure of DIMUS demonstrator system

- nine photocells, three movement detectors, and an events simulation board (by Ansaldo Ricerche) connected to the VDOT32 board;
- four tactile arrays connected to the VDOT32 board;
- two omnidirectional microphones Sennheiser MKE 212 with power supply;
- two cardioid microphones Sennheiser MD 441;
- four omnidirectional microphones PMZ MKR-010.

The software architecture of the DIMUS demonstrator system is shown in figure 3. It can be divided into three parts: "input processing", "functional and control processing" and "output processing". The first one contains the modules dedicated to the acquisition of the three types of data (binary, acoustic, visual), the module for the calibration of the TV cameras and the module for the off-line configuration of the sensors. The "functional and control processing" part contains all the modules useful for the processing tasks: there are modules for the low-level processing of signals and images, for the spatial and temporal modelling of the environment, for the fusion and the interpretation of the data coming from the low-level modules. The system supervisor provides a mean to exchange information between the other modules; moreover it manages alarms, priorities and the shared global resources. The third part contains the modules for the management of the mobile camera and the multiplexer. In the end the user interface manages the interaction with the operator.

3.1 DIMUS demonstrator system: "detection of people in prohibited areas"

This surveillance task requires the capability of a reliable detection of presence of people, also in presence of faulty sensors: in the DIMUS demonstrator system the tunnel has been chosen as prohibited area, both the sidewalk and the portion of tracks.

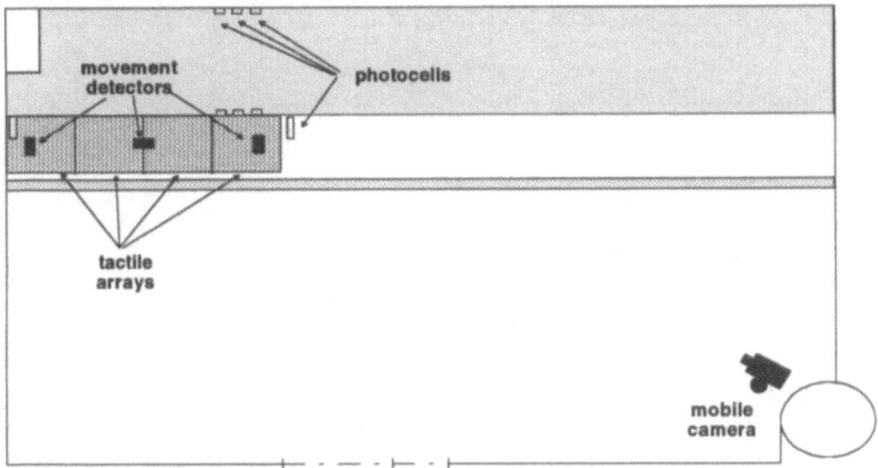

Fig. 4: DIMUS demonstrator system, sensors
used in "detection of people in prohibited areas"

Several classes of binary sensors are used: four tactile arrays, six photocells, and six infrared movement detectors (see figure 4). The four tactile arrays constitutes a tactile carpet in order to detect the presence of people in the sidewalk: the activation sequence of the tactile arrays is used to give to the operator additional information about the walking direction of a person. The three movement detectors hanging from the ceiling support the indication of the movement direction and the persons localisation. Six photocells are used to detect the presence of a person on the tracks. When a presence is detected in the tunnel, on the user interface an appropriate alarm blink, the position of the person is shown on the station map and the mobile camera automatically is pointed toward him.

3.2 DIMUS demonstrator system:
"detection and localisation of anomalous acoustic events"

The detection of anomalous acoustic events in the underground environment is a very important surveillance task, because these events could be related to other more complex dangerous situations. An array of four omnidirectional microphones is used (see figure 5): the processing of acoustic signals coming from these sensors is able to discriminate the type of acoustic event (scream or gun shot) and the location where it come from. When an anomalous acoustic event is detected, on the user interface an appropriate alarm blink, the position of the person is shown on the station map and the mobile camera automatically is pointed towards him.

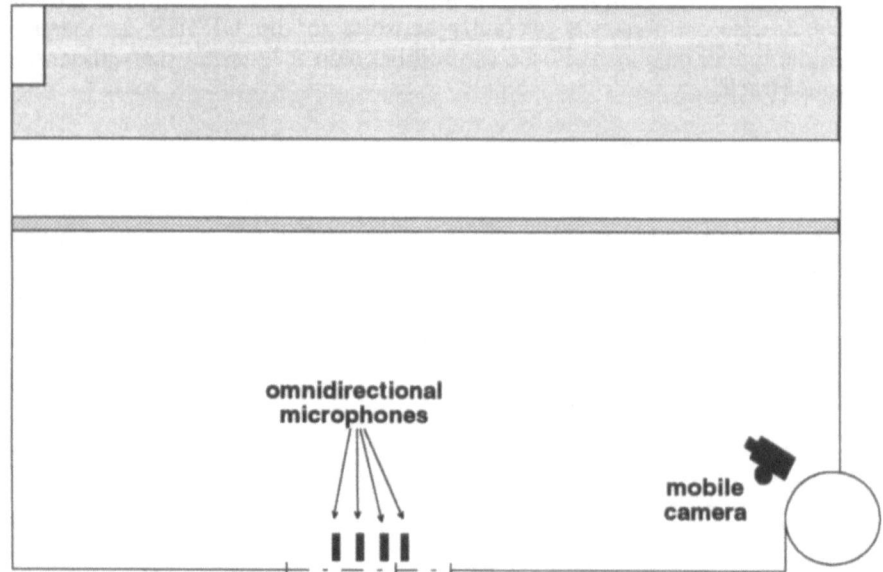

omnidirectional
microphones

mobile
camera

Fig. 5: DIMUS demonstrator system, sensors used in
"detection and localisation of anomalous acoustic events"

3.3 DIMUS demonstrator system:
"detection of people in dangerous zones"

In the DIMUS demonstrator system the border of platform and the tracks represent the dangerous zones. This surveillance task is characterised by the detection of two different dangerous situations:
- presence of people in the area beyond the yellow line and near the tracks;
- presence of people fallen onto the tracks.

Both these situations are related to the presence or absence of the train in the station: the corresponding alarms will be inhibited during the train-stop.

The TV camera (see figure 6) is used for two controlling actions: the tracks area and the region close to the platform boundary. The chosen method is to define two regions of interest for the two above mentioned surveillance actions in which moving objects can be detected by means of low level image processing algorithms. A photocells curtain is placed between the two surveilled regions in order to create a virtual separation for controlling who is moving towards the tracks. This sensor array takes advantage of the fast response time allowing a quick evaluation of the presence of offending objects. The photocell barrier is organised as a system with interposed beams for determining, approximately, the intruder dimensions. A set of movement detectors, hanging from the underground ceiling in the track area, can be considered as a third sensor system that co-operates with the others in checking the presence of people in the dangerous zone: as the field of action is elliptical, they can cover both the tracks area and the platform boundary and they can determine an approximate position of the offending object along the platform. When a presence is detected, on the user interface an appropriate alarm blink, the position of the person is shown on the station map and the mobile camera automatically is pointed towards it.

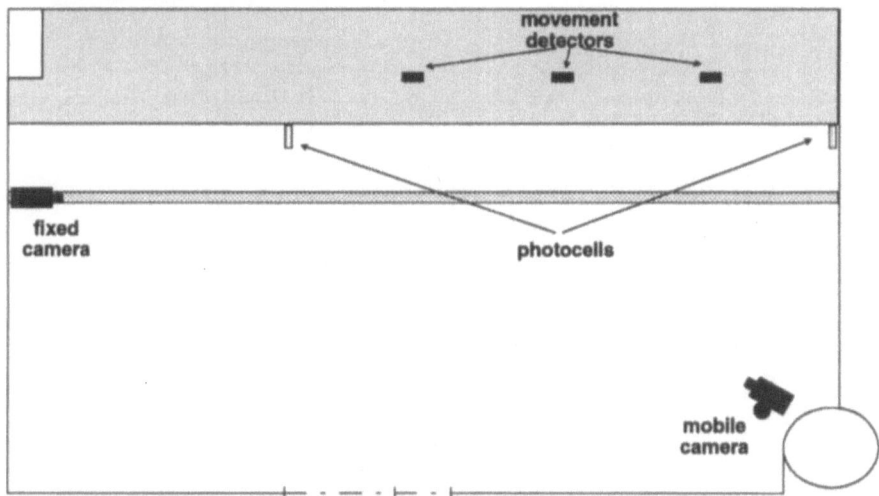

Fig. 6: DIMUS demonstrator system, sensors
used in "detection of people in dangerous zones"

3.4 DIMUS demonstrator system: "evaluation of level of crowding"

In an underground station, can be useful to distinguish between "usual" overcrowding and "anomalous" overcrowding events. An "usual" overcrowding event is considered the situation when too many people are in the station waiting for the train (this can be caused by peak traffic hours, special occasion, etc.). In this case, people is uniformly displaced all over the station, and resulting event is therefore defined as "absolute overcrowding". An "anomalous" overcrowding event is considered the situation when too many people gather in an area of the station for unknown reasons (this can be caused by people aiding hurt persons, people fighting, etc.). In this case, people is not uniformly displaced all over the station (on the contrary, vast people free station sub areas should be detectable), and resulting event is defined as "relative overcrowding". The evaluation of level of crowding in the DIMUS demonstrator system is characterised by two complementary aspects:

- absolute/relative overcrowding conditions are detected and urgency modulated asynchronous alarms are generated: overcrowding alarm buttons blink and (in case of relative overcrowding)the sub area where the event has been detected is enlightened;

• evaluation of crowding is cyclically performed, in order to estimate the number of people in the station, with no alarm generation.

Four crowding levels will be discriminates in the DIMUS demonstrator system:
- no people
- a few people (1 to 3 persons in the simulated underground platform)
- many people (4 to 10 persons in the simulated underground platform)
- too many people, that is overcrowding (more than 10 persons in the simulated underground platform)

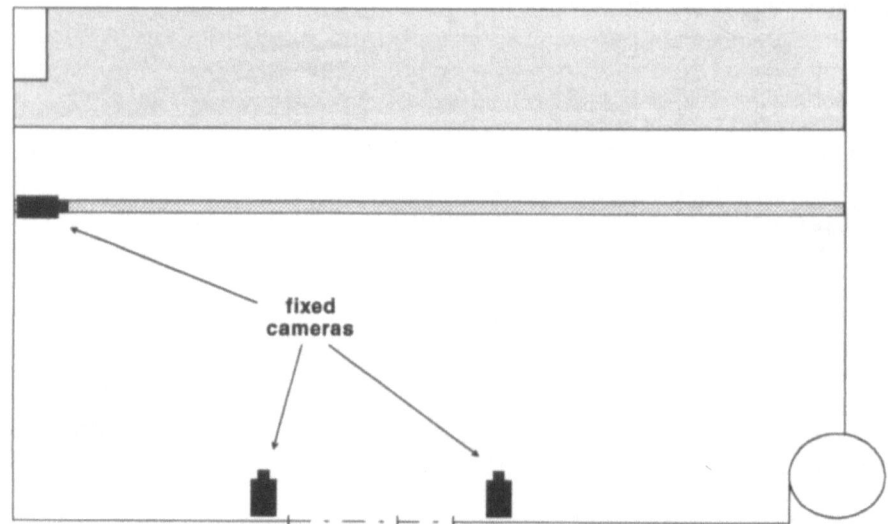

Fig. 7: DIMUS demonstrator system, sensor used in "evaluation of level of crowding"

For the development of this surveillance task TV cameras (see. figure 7) and different technologies have been used: image processing, neural networks, knowledge based systems.

3.5 The management of concurrent events in the DIMUS demonstrator system

The DIMUS demonstrator system handles the concurrency:
- concurrency of alarms: one alarm occurs while another one is being processed by the system;
- concurrency of use of global resourses: one module (or the user) tries to get access to a global resource (the mobile camera) while it is already under control of another module.

These two aspects of concurrency are handled by the system supervisor using one single mechanism, which consists to arrange the alarms or requests coming into the supervisor by urgency. The DIMUS demonstrator system is able to show its behaviour in presence of the following situations of concurrency:
- low urgency alarm followed by a high urgency one;
- high urgency alarm followed by a low urgency one;
- concurrent access to the mobile camera;
- simulated events (to test the robustness of the system).

4. Future developments of the DIMUS project

As already said, a second phase of the DIMUS project is foreseen, with the goal of the installation of the demonstrator system in a station of Genoa Underground, in order to test the functionalities and the performances in real working environment. This phase of the project will be carried out with the collaboration of Azienda Municipalizzata Trasporti of Genoa and Ansaldo Trasporti, which will perform DIMUS as a future product. Moreover, the DIMUS project could implement other surveillance functionalities, (arisen from contacts with Italian underground Companies) as the detection of abnormal people flow, the detection of unattended objects or the detection and the tracking of moving people.

Acknowledgements

The authors thank all the members of the DIMUS consortium for their contribution in the development of this work. Special thanks to Ansaldo Trasporti and Azienda Municipalizzata Trasporti of Genoa for the cooperation in the definition of requirements of the DIMUS application in order to obtain a future product.

References

1. ESPRIT P5345 DIMUS, "Technical Annex", October 1990
2. ESPRIT P5345 DIMUS, "First Design Specification Report", September 1991
3. ESPRIT P5345 DIMUS, "Integrated Software Documentation", September 1992
4. ESPRIT P5345 DIMUS, "Intermediate Demo Report", September 1992

A Reflex-Based Approach to Fusion of Visual Data

A. Bozzoli, M. Rossi, R. Barbò, B. Caprile, G. Carlevaro
Istituto per la Ricerca Scientifica e Tecnologica
Località Pantè di Povo, I-38050 Trento

Abstract

A system able to estimate in real-time the crowding level of an indoor environment is described. It receives visual information from cameras and it is based on a trainable Hyper Basis Functions network (HBF) that learns to map from a series of "examples" provided by a human operator to map visual inputs into crowding levels.

In this paper we present the preprocessing algorithms used to extract special features from gray levels images and a brief account of HBF technique. Experimental results obtained in real-life situations are finally reported and discussed[1].

1. Introduction

Capability of making fast the estimate of the number of people present in a given environment is a highly desirable feature for any surveillance system, especially for those designed to work in situations in which high levels of crowding are potentially dangerous [4, 7]. The present note is devoted to the description of the crowding evaluation system we have been developed during the last year, which is implemented in a reflexive[2] fashion.

From a functional point of view the system consists of two modules: a Preprocessing Module which transforms raw visual data into a numerical vector describing meaningful features; an Fusion Evaluation Module consisting in a HBF network [8, 9, 10] mapping the input provided by the Preprocessing Module to the corresponding crowding level.

The organization of the paper is as follows: section 2 describes the actual task and outlines our approach; in section 3 we present the structure of the

---

[1] Research described here is part of Espri Project P5345 DIMUS [1] aimed at the development of an information processing system capable of supporting the activity of a human operator in the surveillance of underground station environment.

[2] The evaluation of the crowding level is performed at a high rate, and that the system is provided with no "explicit representation" of the world.

system, the algorithms used in the low-level preprocessing and introduction to the HBF networks; section 4 summarizes experiments and results and section 5 contains some final considerations and directions for future work.

2. The Problem

Let us state first the application to be solved: given a certain set of cameras, adequately displaced inside a room (e.g. an underground station) we have to estimate how many people are present in the room. We have considered four levels of crowding: "1", "2", "3", "4", associated to the presence of, respectively, "one person", "few people", "many people" and "overcrowding".

The underlying idea is to use the movement detection techniques [6] to get useful information related to the overcrowding level of the inspected scene. The steps towards the solution consist in finding a set of features, based on visual inputs, capturing in a compact way the relevant information and mapping them into the corresponding crowding level.

3. The System

The system consists of two main modules (see figure 1): the Low-Level module [3] and the Fusion Evaluation Module [5]. The first one is devoted to the pre-processing of images and feature extraction; the second one consists of the HBF network to determine the current level of crowding.

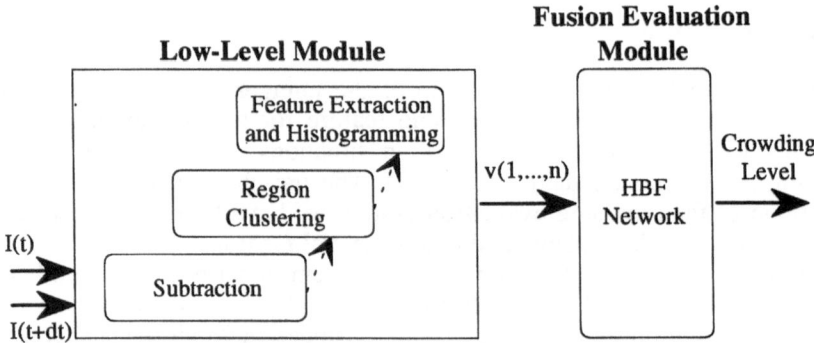

Fig. 1: The overall system architecture: given two images I(t) and I(t + Δt) the crowding level is computed.

3.1 The Low-Level Module

The role of the low level module is to extract from sequences of images information relevant in a format suitable to be used as input to the Fusion

Evaluation Module. The preprocessing procedure adopted rests upon the assumption that a reasonable fraction of the people in the room keeps on moving and the changes in the scene are different in the case of few or many people. This depends on the mutual occlusions among many moving persons causing a "fragmentation" of the motion pattern.

The whole process has been made up of the following procedural steps: images subtraction, blob coloring, blob feature extraction, and blob feature histogramming. Two are the main approaches to the detection of temporal changes on images that we have examined considering the variation in the gray level intensity function: subtraction of an acquired image against a stored reference one (representing the background of the scene), and subtraction between two (temporally) successive images. This can be written as:

$$\Delta I = I_g(x,y,t_j) - I_g(x,y,t_i) \tag{1}$$

where $t_j = t_i + \Delta t$ with Δt = temporal distance in number of frames.

These two approaches might be used as consecutive steps of the same procedure: the first one can be used as an alerting module, able to trigger the second one whenever there are people in the observed scene. The second method is going to be more robust in respect to illumination conditions, while the first one does not require that people are moving to be detected.

These methods have been considered to get a binary map from gray levels images representing moving objects or, in general, changes in the scene. According to the (1) the pixel value is set to "1" if the absolute value of the difference of the gray level intensity function between corresponding pixels exceeds a given threshold whilst is set to "0" if the opposite case occurs:

$$I_b(x,y,t_i) = \left\langle \begin{array}{lll} 0 & if & |\Delta I| \leq THR \\ 1 & if & |\Delta I| > THR \end{array} \right. \tag{2}$$

Although the best way in attributing the value of the corresponding pixel of the binary map is to take into account the pixels neighborhood information whenever the absolute difference value lies in a given range [6], the adoption of this simplified method is justified by elaboration time constraints. The pixels marked "1" are considered the result of object motion, and they are called moved pixels.

A clustering technique mapping individual pixels into a region has been evaluated. We have found particularly appealing the blob coloring algorithm [2]: a local technique where the pixels are placed in a non-simply 4-connected regions (in the following called blobs) on the basis of the near neighbourhood properties. Given a binary image, this algorithm assigns a different color label to each blob. The image is scanned left to right and top to bottom by using a given L-shaped template [2]. This template allows to partition the image into equivalence classes of pixels to which the same color has been assigned.

The obtained color map allows us to extract some blobs features, such as: the area, the axes of the bounding box, and the shape factor (compactness).

Considering that our Fusion Evaluation Module requires that selected features be mapped into a linear space of a given dimension, it seems that the histogram technique is the most straightforward to be implemented. The criterion was to analyse the histogram shape using the modality of the data for each extracted features and for each different crowding level with the aim to define the correlation between the features space and the real life situation. Many experiments have been done (see section 4) to find the optimal feature and a good histogram partition to get n-tuples of fixed length that, used as input of the Fusion Evaluation Module, could discriminate as best as possible the different crowding levels in the inspected scene.

3.2 The Fusion Evaluation Module

The evaluation of the crowding level can be seen as a classification problem. Let us assume that the output of the preprocessing phase consist of a n-tuple of numbers embodying relevant information about the scene (see section 3.1). The problem of building a crowding level classifier can therefore be made equivalent to the problem of finding a function

$$g : R^n \rightarrow L$$

(where L is the subset of the Integers representing the totality of the crowding levels) such that:

$$g(x) = l; \quad \forall x \in X_l, \tag{3}$$

where X_l is the set of possible outputs of the preprocessing module when the number of people present in the room corresponds to the l-th crowding level. Of course, this is quite an ideal statement of the problem, and due to the presence of noise the function g may not even exist. Our approach to the problem can therefore be summarized as it follows:

- consider first a set of pairs of the form $D \equiv \{(x_i, l_i)\}_{i=1}^{\cdots}$. The set D can be seen as a series of samplings of the hitherto unknown mapping from the totality of sensorial inputs[3] to L that we are seeking for;
- find an approximation to data D in terms of a Hyper Basis Functions network.

Once that a "sufficiently good" approximation to data D is found, we will say that our system *has learnt* how to approximate information that examples D contained.

Before entering a brief description of the HBF technique, a remark is worth doing: Hyper Basis Functions networks compute *continuous* functions, whereas classifying mappings we are looking for are intrinsically discontinuous.

Some *discretizing* procedure for the output of the HBF networks has therefore to be devised in order to map real numbers into the subset of integers representing the crowding levels. Our choice was to round the output of the network to the nearest integer value in the set L.

---

[3] With the term "sensorial input", we will refer here to an output n-tuple of the preprocessing module.

3.2.1 Hyper Basis Functions Networks

Hyper Basis Functions networks can be regarded as special cases of a wider class of networks, called *Regularization Networks* recently introduced by Poggio and Girosi as a general approximation technique that can be used in problems of learning from examples. A regularization network realizes functions of the form:

$$f(x) = \sum_{\alpha=1} c_\alpha \, G(\| x - t_\alpha \|_W) \tag{4}$$

where

$$\| x - t_\alpha \|_W = (x - t_\alpha)^T W^T W (x - t_\alpha)$$

and W is a square matrix that has to be estimated along with the coefficients c_α and the "centers" t_α. Here G is intended to be any (conditionally) positive definite function, such as the Gaussian or the Multiquadric [11, 12].
Given a sparse data set:

$$D \equiv \{x_i, y_i) \in R^d \times R\}_{i=1}^M$$

we will say that a good approximation to D has been found in the form (4), when values of parameters $\{c_a\}_{\alpha=1}^n$, $\{t_a\}_{\alpha=1}^n$, and W have been determined which minimize the square error function

$$E(c_1, ..., c_n, t_1, ..., t_n, W) = \sum_{i=1}^M (f(x_i) - y_i)^2 \tag{5}$$

Two characteristics of the HBF are worth mentioning. The first one consists in the fact that optimal choice of the centers corresponds to best (optimal) clustering of data $\{(x_i, y_i)\}_{i=1}^{...}$, since every center can be thought as representing a template to which network inputs will be compared. The second one relates to the meaning of the weights W: any instance of W corresponds to a particular choice for the metric of the input space, and distances among inputs are accordingly evaluated. Optimal values for the entries of matrix W will therefore correspond to a linear transformation yielding the "most natural" set of coordinates for the input space [8].

Finding optimal values for parameters which functional (5) depends on is a hard minimization problem. In many cases of interest, its complexity can nevertheless be dramatically reduced by establishing extra constraints upon values of the arguments of E (typical cases consist in imposing that W be diagonal, and/or that centers t_α be fixed); yet analytical solutions stand out of reach, and numerical minimization methods have therefore to be applied.

Our experience has clearly indicated that - due to the presence of several many local minima - classical minimization methods such as Gradient Descent or Conjugated Gradient do not perform particularly well in the task of minimizing functional (5). Seeking for a reasonable compromise among effectiveness, computational cost and efficiency of the minimization process, we have therefore been led to consider nondeterministic methods as possible ways to tackle our problem. A nondeterministic optimization method recently proposed by Caprile and Girosi [13] appeared to be fairly promising in our cases - having it already been tested on tasks very similar to the one we were engaged to solve. Essentially, the algorithm consists in a random walk across the parameters space, whose step size along different axes is continuously adapted according to whether the previous step has led or not to a decrease in the value of functional (5). Beside the level of performances it has shown, one of the features of the method we have found most appealing in the fact that no fine tuning of parameters which the minimization process depends on has ever been required - quality of the results we have obtained having been pretty constant over a wide range of possible choices.

4. Experiments and Results

Following the proposed pre-processing schema (see figure 1), some experiments have been performed to select the low-level procedures giving the most promising histograms, which enable the Fusion Evaluation Module to discriminate the different crowding levels.

Figure 2 shows the binary maps obtained from the image subtraction algorithm whilst in figure 3 a typical blob histogram is given.

Fig. 2: Difference images. From left to right and from top to bottom: one person, few people, many people and overcrowding.

The area of the blobs, figure 4, seems to be the best choice compared to the others features analysed. Indeed, for each binary map, the histograms of the bounding box axes and of the compactness of the blobs are too sparse, showing values closely concentrated nearby the origin.

Analysing the shape of the cumulative histograms of each sequence of binary maps associated to a given crowding level and evaluating the first and second order statistical parameters (average, mode, median, variance and coefficient of variation) a significant partition of the features histogram has been chosen (see figure 3).

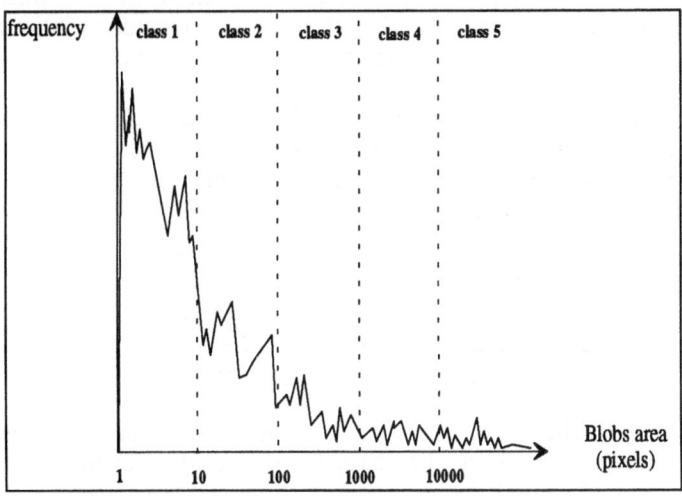

Fig. 3: Partitioned histogram of the blob area.

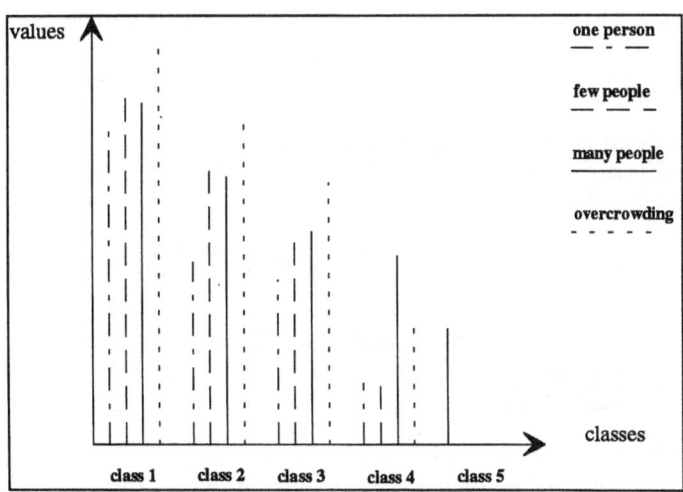

Fig. 4: Histograms of the blob area of the difference image shown in the figure 2.

The five classes, characterized by two thresholds and defining the number of blobs whose areas belong to the correspondent range, form the input vector to HBF network:

$$v = v(s_1, s_2, s_3, s_4) \tag{6}$$

with s_i = *sum* of the blobs belonging to class i.

The first step, in applying the HBF network, was to collect a reasonable number of examples for each crowding level. What turned out was that two hundreds of data is a sufficiently representative sample for each single level. Once that the learning set has been collected, the network was trained. The behaviour of the network is conditioned by two parameters: the temporal distance between two images Δt, and the number of fixed centers (see section 3.2.1).

In the first case we have considered three values of Δt between two images: 1, 8 and 25 (in number of frames). Tests were done using one hundred images for each crowding level and results are given in table 1 showing that 8 is the optimal value among those considered.

D(t)	n. centers	results
1	50	76%
8	20	96%
25	20	74%

Table 1: The table shows, for each temporal distance considered, the best percentual of correct evaluations changing the numbers of network centers. The distance 8 gives the best result detailed in the next table.

This is in agreement with the following observation: the difference between images at distance "1" is of very little meaning unless people move fairly quickly, whilst a temporal distance of 25 frames may result in too uniform a shape of the features histogram - making the classification of the classes an arduous task to accomplish.

In the second case trying different values of fixed centers chosen from 10, 20, 50, and 100 (see section 3.2.1) we have found 20 the best result for the HBF network (see table 2).

	CL1	CL2	CL3	CL4
CL1	98	2	-	-
CL2	12	88	-	-
CL3	-	-	100	-
CL4	-	-	2	98

Table 2: Matrix of confusion of the results. The diagonal reports the number of the correct evaluations, whose average on all the crowding levels is about 96%. The main faults occur in discriminating levels 1 and 2.

5. Conclusions

A system to evaluate the crowding level of an environment has been described. The low-level layer has proven to be able to provide the information the Fusion Evaluation Module needs in a fast and reliable way. Experimental results - though preliminary - seem to be promising for the further development of our research. Better and more reliable results may be obtained by seeking for more stable and significant low-level features, and in a more effective training procedure for the HBF network (in particular for what concerns the choice of the centers).

Algorithms we have described have been implemented in C language on a SUN Sparkstation 2 linked to the Signum S151 image processing system. Time required for a single evaluation of the crowding level, based on images at the resolution of 512x512 pixels, is smaller than 5 seconds, and the training of the network requires less than 30 minutes. These numbers show that a well engineered release of the system, taking advantage of fairly standard hardware components, may perform several evaluations per second - therefore approaching real time performances.

References

1. Various Authors, "First Design Specification Report", Esprit Technical Report, DIMUS-D1T1200, 1991.
2. Ballard D. H. and Brown C. M., "Computer Vision", Prentice Hall, Englewood Cliffs, 1982.
3. Barabino G., Barbò R., Bozzoli A., Bisio G. M., Peri M., Regazzoni C. and Vernazza G., "Low Level Algorithms", Esprit Technical Report, DIMUS-D1T3300, 1992.
4. Brachet J. C., "Fall Detection System in a Metro Line".
5. Caprile B. and Rossi M., "Fusion Algorithm Implementation", Esprit Technical Report, DIMUS-D1T4100, 1992.
6. Netravali A. N. and Robbins J. D., "Motion-Compensated Television Coding", The Bell System Technical Journal, vol. 58, No. 3, pp. 631-670, March 1979.
7. Sasama H. and Ukai M., "Application of Image Processing of Railways", QR of RTRI, vol. 30, pp. 74-81, Japan 1989.
8. Poggio T. and Girosi F., "Extension of a Theory of Networks for Approximation and Learning: Dimensionality Reduction and Clustering", Proceedings Image Understanding Workshop, Pittsburgh, Pennsylvania, pp. 597-603, September 11-13, 1990.
9. Poggio T. and Girosi F., "Networks for Approximation and Learning", Proceedings of the IEEE, vol. 78, No. 9, September 1990.
10. Poggio T. and Girosi F., "A Theory of Networks for Learning", Science, vol. 247, pp. 978-982, 1990.
11. Poggio T. and Girosi F., "A Theory of Networks for Approximation and Learning. A. I. Memo No. 1140, Artificial Intelligence Laboratory, Massachusetts Institute of Technology, 1989.

12. Micchelli C. A., "Interpolation of Scattered Data: Distance matrices and Conditionally positive definite functions", Constr. Approx., vol. 2, pp. 11-22, 1986.
13. Caprile B. and Girosi F., "A Nondeterministic minimization Algorithm. A. I. Memo 1254, Artificial Intelligence Laboratory, M.I.T., Cambridge, MA, September 1990.

Sensor Fusion in a Peg-In-Hole Operation with a Fuzzy Control Approach

Jianwei Zhang, Jörg Raczkowsky
Institute for Real-Time Computer Systems and Robotics,
Department of Computer Science, University of Karlsruhe, Germany

Abstract

An advanced robot system should be able to make independent decisions with the aid of diverse sensor information. Fuzzy logic is an appropriate tool to model and describe human experience and intuitive knowledge for decision-making. In this paper, we present a realization of a peg-in-hole operation with a fuzzy logic control approach. The sensor data of a force/torque sensor and a mini-camera vision system are integrated, and the fused information is used in order to decide the motion of the robot gripper for performing the insert operation. The information of these two sensors is used complementary to support the assembly process. Every data of a sensor is a weighted evidential information which influences the fuzzy inference. Additional sensors will be integrated in the next project phase. The above described research work is supported by the DFG German Research Council under contract SPP-RE489/22-2 and carried out at the Institute for Real-time Computer Systems and Robotics of the University of Karlsruhe.

1. Problem Description

KAMRO (KArlsruhe Mobile RObot) is an autonomous robot system, which is now being developed at the Institute for Real-Time Computer Systems and Robotics (IPR) of the University of Karlsruhe (see figure 1). It serves as a test-bed for implementing an intelligent, fault-tolerant robot control system which should be able to operate in an industrial environment, [1].

There are three main components of KAMRO: the mobile vehicle, two manipulators and the sensory system. Among them the two manipulators perform assembly tasks, like grasping, insertion, screwing, etc. The sensory system consists of ultrasonic sensors, vision systems and a force/torque sensor.

In order to perform assembly tasks autonomously, the KAMRO-system should be able to realize a set of basic functions called "Explicit Elementary Eperations" (EEO) like *Transfer, Finemotion, Grasp/Detach, Join/Disjoin*, etc. [2]. A typical problem in EEO *Join* is the "*peg-in-hole*" (also called *Pin Insertion*). The gripper of a robot arm grasps a peg and puts it into a hole whose size is relatively small. Generally, two consecutive phases are needed to place a peg into a hole:

1. *search of the hole* , taking the peg to the position over the mouth of the hole
2. *push-into* - pushing the peg in until end of the hole is reached (see also [3] for a description of the peg-in-hole operation).

We will discuss a solution of the peg-in-hole problem with a fuzzy logic control (also called fuzzy control) approach.

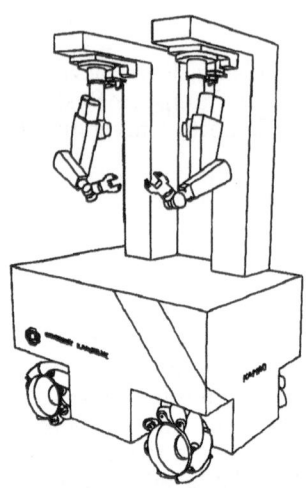

Fig. 1: An autonomous robot system KAMRO

KAMRO-system is assumed to work in an unstructured, dynamic environment. Therefore, there exist many uncertainty factors which make the normal off-line programming method unsuitable for the peg-in-hole task. In the experiment with KAMRO [1], the peg-in-hole operation was realized in the following steps:

1) determine the position of the hole according to the information from a static overhead camera system;
2) grasp the peg with the gripper and move to this position using the EEO command *Finemotion* in the world coordinate system;
3) perform the *push-into* task.

Unfortunately, in some real tests it is found that after steps 1 and 2, the peg is not located precisely over the mouth of the hole, but on the surface near the hole. This is caused by the following reasons:

a) the limited resolution of the vision system;
b) the difficulty for the robot arm on a mobile vehicle to calibrate in the world coordinate system; and
c) the inaccuracy of the arm. It is thus necessary to develop a strategy for the robot to find the correct hole position considering all the uncertainty factors.

Fuzzy set and fuzzy logic theory, founded by Zadeh in 1965, [4] [5], provides an appropriate mechanism to describe, model and infer with fuzzy concepts, intuitive knowledge and control strategies. Recently, fuzzy logic is neither constrained in the theory research nor limited in laboratories any more. It has been put into practice, especially in the area of control. Fuzzy control [6] is linguistic control.

The input and output variables are treated as *linguistic variables*, such as distance, speed, etc. Each linguistic variable is represented by a set of *linguistic terms* interpreting fuzzy sets, e.g. linguistic terms of *"distance"* could be"near"(N), "middle" (M), and "far" (F). The input and output variables are treated as *linguistic variables*, such as distance, speed, etc. Each linguistic variable is represented by a set of *linguistic terms* interpreting fuzzy sets, e.g. linguistic terms of *"distance"* could be"near"(N), "middle" (M), and "far" (F).

A *membership function* μ is used to describe the degree with which an input value of a variable belongs to a fuzzy set. A fuzzy controller consists of four components: fuzzification, inference engine, rule/data base, and defuzzification, see [6] for a summary of fuzzy control. In our work we use the max-min inference, and the COG (Centre of Gravity) defuzzification method.

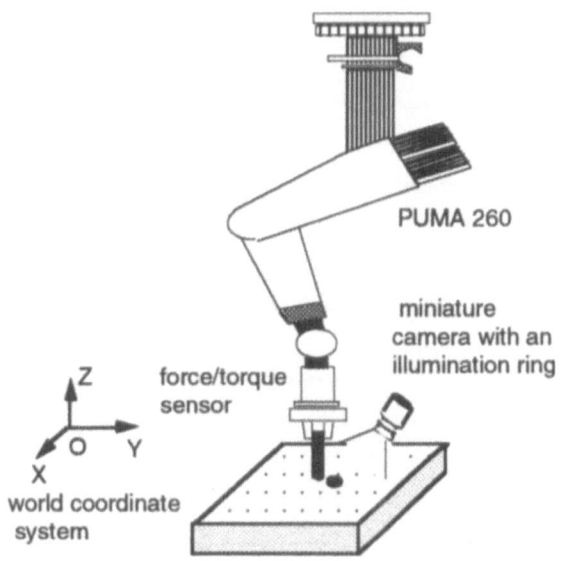

Fig. 2: The system set-up for the peg-in-hole operation

2. Strategy for Searching the Hole Based on Vision and Force/Torque Sensors

In the gripper of the PUMA 260 manipulator, a force/torque sensor is integrated which measures three forces and three torques on the wrist. Besides the overhead camera, a miniature camera (16mm CCD video camera) system, on which an illumination device is mounted, is being developed at IPR. It is intended to be fixed on the wrist of the robot arm or grasped by the gripper. In the test phase of our experiment, the mini-camera is fixed near the hole, having an appropriate angle with the peg to be inserted so that the camera can take clear images of the peg and the hole in its view field. Figure 2 shows the system set-up for the experiment. Based on the vision and force/torque sensor information, the following strategy is proposed to control the robot arm to find the hole.

In step 1 and 2, the robot gripper takes the peg and moves towards the assumed hole position until a contact with an object is detected by the force/torque sensor. The detection of the peg locating in the hole can be done by slightly moving the peg horizontally and check if forces against the moving direction exist. If the peg does not locate in the hole, then a position correction algorithm is used to control the robot to find the hole. This algorithm is schematically illustrated in figure 3.

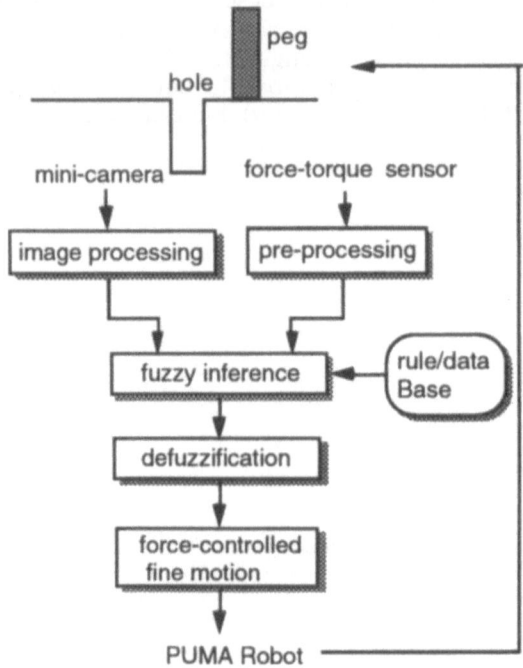

Fig. 3: Correction of position error using fuzzy control approach

At the beginning of each correction step, the mini-camera and the force/torque sensor are activated. A camera image is taken and processed so that information for position correction can be extracted. After the pre-processing of the force/torque sensor, three forces and three torques are available. These sensor data serve as the inputs for the fuzzy inference module. Both the input and the control variables are interpreted as linguistic terms and the correction strategies are modelled as fuzzy control rules, which are stored in the rule/data base. After the defuzzification, the numerical values of the correction vector in the world coordinate system are evaluated. They are used for executing the force-controlled fine motion to perform the correction.

3. Information Extraction from Images of the Mini-Camera

For simplicity, the mini-camera is located near the peg and the hole so that its view field covers only part of the peg and the hole. A camera image is first pre-processed through digitization and segmentation (only binary images are considered here). Then in the 2-D image coordinate system, there are only two cases for the image interpretation to be distinguished, two separate objects, and two overlapping objects (see figure 4 for two corresponding images).

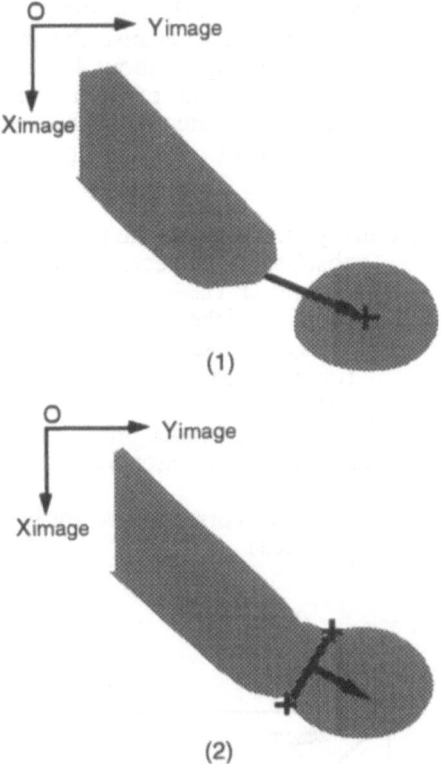

Fig. 4: Two images from the mini-camera

In the first case it is necessary to identify which object is the hole and which is the peg. This is done through the following steps:

i) First, the centres of gravity (COG) of the two objects are computed and eight test radial lines are extended from each COG until they have intersections with the object edge.

ii) If the eight segments have almost the same length, this object is regarded as the hole. The other object is thus the peg. A point from that edge of the peg image is selected which is closest to the hole image.

iii) The vector from this point to the COG of the hole is then the correction vector in the image coordinate system (see part 1 of figure 4).

In the second case the separation line of the peg and the hole must first be decided. Along the edge of the segmented object, two points locating at the most concave positions are selected. A segment connects these two points. Then the correction vector is regarded as the line perpendicular to this segment with the same length (see part 2 of figure 4).

4. Development of Fuzzy Control Rules

After information extraction, a correction vector **T** in the image coordinate system X_{image}-O-Y_{image} can be computed. To correct the position of the robot gripper, **T** must be transformed to the world coordinate system. Here only an approximate transformation is proposed because the mini-camera is located obliquely. If the camera is fixed with angles about 45° to the X, Y and Z axes, the world coordinate system is then about 45° rotation from the image coordinate system, see figure 5. An auxiliary coordinate system X_{image}'-O-Y_{image}' is created by rotating the X_{image}-O-Y_{image} in 45°, which approximates a scaled word coordinate system.

The input variables for the fuzzy inference are the following:
- *Distance_Image_X* :   abscissa of the vector **T** in the coordinate system X_{image}'-O-Y_{image}';
- *Distance_Image_Y*:   ordinate of the vector **T** in the coordinate system X_{image}'-O-Y_{image}';
- *Force_Z:*   force in Z axis of the world coordinate system. It serves for monitoring if the hole is found.

The two output variables of the fuzzy inference are:
- *Correction_X*: : component of the correction vector in X-axis in the world coordinate system X_{world}-O-Y_{world}.
- *Correction_Y*: : component of the correction vector in Y-axis in the world coordinate system X_{world}-O-Y_{world}.

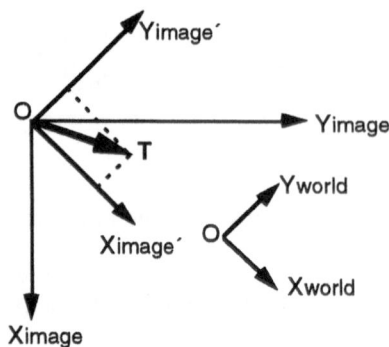

Fig. 5: Transformation from image to world coordinate system

The linguistic terms of these variables can be defined as the following:
Distance_Image_X = { NB, NM, NS, Z, PS, PM, PB}, see part (a) of figure 6.
Distance_Image_Y = { NB, NM, NS, Z, PS, PM, PB},
Force_Z = { Z, N }, see part (b) of figure 6.

For the output variables:
Correction_X = { NB, NM, NS, Z, PS, PM, PB }, and
Correction_Y = { NB, NM, NS, Z, PS, PM, PB },
which are defined similarly as in part (a) of figure 6. In the phase of *search of the hole*, the gripper position is corrected linearly.

Control rules look like the following expressions:

> **if** *Distance_Image_X* = PB and *Force_Z* = N **then** *Correction_X* = PB
> **if** *Distance_Image_Y* = PM and *Force_Z* = N **then** *Correction_X* = PM
>
> ...
>
> **if** *Force_Z* = Z **then** *Correction_X* = Z and *Correction_Y* = Z

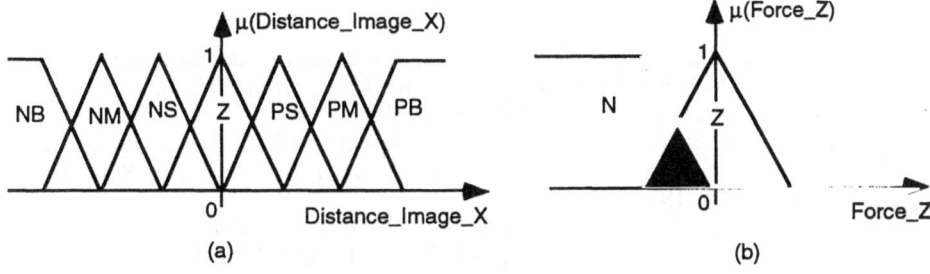

(a) (b)

Fig. 6: Linguistic terms of the two input variables

5. Discussion

For the mini-camera, it is desirable that the peg does not hide the hole. So the control rules can be modified correspondingly by damping the correction in the direction to the camera. We are considering to let the other robot arm grasp the mini-camera (thus becoming a *hand-eye*) and monitor the insertion phase in a mobile manner. Then the camera can be positioned at any suitable location. For example, if the camera has got no certain information for the correction, it can be moved by the robot arm to the opposite side to take a second image. The information from both of these two images can be input to the fuzzy inference and then synthesized to make a more reliable correction.

Another interesting work is to apply the fuzzy control to the *push-into* phase, which is especially meaningful if the peg and the hole have a tight clearance. The mini-camera vision system can detect the error in the orientation of the gripper from outside, while the three force and three torque values can be utilized to determine the orientation and position errors from the inner side. The forces and torques will be interpreted more finely to describe their influences on the correction. In this way, the data from vision and force/torque sensors can be effectively integrated.

6. Conclusions

Fuzzy control approach provides a mechanism to utilize uncertain or imprecise data. It is very effective if the system consists of several sensor inputs or intermediate process states which work together to influence a common control action. In the manipulator motion control, the position information and the force/torque sensor values influence the insertion action. In this application, the fuzzy control approach provides an intuitive, easy understandable and effective tool for developing a control software. Instead of "hard" and sophisticated algorithms, it also gives a satisfactory alternative solution. For many tasks which an intelligent robot system should fulfil, we as human beings have accumulated redundant experience and intuition from our daily life. This knowledge can be best summarised and modelled by fuzzy control rules and realized by a fuzzy controller.

References

1. Hörmann A., Meier W., and Schloen J., "A Control Architecture for an Advanced Fault-Tolerant Robot System", *Int. Conf. on Intelligent Autonomous System IAS*. 2, Amsterdam, 1989.
2. Cheng X., Kappey D., and Schloen J., "Elements of an Advanced Robot Control System for Assembly Tasks", *5th Int. Conf. on Advanced Robotics*, Pisa, pp. 411-416, 1991.
3. Inoue H., "Force Feedback in Precise Assembly Tasks", in *Artificial Intelligence: An MIT Perspective*, vol. 2, Cambridge, MA: MIT Press, pp. 219-241, 1981.
4. Zadeh, L. A., "Fuzzy Sets", *Information and Control* 8, pp. 338-353, 1965.
5. Zadeh, L. A., "Outline of a New Approach to the Analysis Complex Systems and Decision Processes", *IEEE Trans. on System, Man and Cybernetics*, Vol. SMC-3, No. 1,,pp. 28-44, January 1973.
6. Lee, C. C., "Fuzzy Logic in Control Systems: Fuzzy Logic Controller, Part I and II", *IEEE Trans. on Systems, Man and Cybernetics*, Vol. 20, No. 2, pp. 404-435, March/April 1990.

Fuzzy Logic Techniques for Sensor Fusion in Real-Time Expert Systems

J. A. Aguilar-Crespo, J. M. Domínguez, E. de Pablo, and X. Alamán
Instituto de Ingeniería del Conocimiento
Universidad Autónoma de Madrid, Spain

Abstract

Sensor data fusion is one of the main problems when developing on-line real-time expert systems. During the development of the MIP project several fuzzy logic techniques have beenbuilt up to help with this problem. MIP [1] is a real-time expert system for assistance to petrochemical processes, deployed and in production in a petrochemical plant of INH-REPSOL in Tarragona, Spain. The techniques presented in this paper ascertain confidence values to each sensor measurement. These techniques are currently in production in the plant.

1. Problem Introduction

The development of on-line real-time expert systems involve many problems, mainly within the area of industrial process control. One of them is the low reliability of the incoming data: wrong measurements are frequently obtained from sensors because of various technical reasons (spurious readings, malfunctions, degradations, etc.). Human experts deal with this problem by checking the validity of all the measurements before taking any decision. The traditional solution is to use different sensors for measuring a certain property, and then to perform a fusion of the data from these sensors in form of a final estimate of the property. So, an expert system has to consider the data fusion as an important task: inconsistent input data has to be rapidly detected and recognized.

The MIP system is based on a blackboard architecture; there are several modules (i.e. the expert module, the simulator module, the communications module, the user interface module, etc.) that work simultaneously and share information through the blackboard. As opposed to including the validation as one of the responsibilities of the expert module, an additional module monitors on-line the incoming data for suspicious behavior and assigns *confidence values* (values in the interval [0,1]; the greater the measurement reliability is, the greater the confidence value) to each measurement. This specialist module has been implemented by means of a fuzzy logic shell that generates C code. So, the fuzzy logic module was easily and quickly integrated into the system.

Why use fuzzy logic techniques to handle this problem? There are several reasons:

- There exist fuzzy logic inference engines in the market that meet the runtime efficiency conditions required by real-time applications. Hardware implementations are also available in case of stricter runtime requirements.
- Fuzzy logic systems are easy to tune in real applications. This aspect is extremely important for systems that must be actually installed and put into production.
- Fuzzy logic reasoning is adequate to handle imprecision in the measurements, e.g., the concepts of "two similar measurements" or "a steady measurement" are basically fuzzy.
- It is easy to use these fuzzy concepts in rules and linguistic descriptions that provide the confidence deserved by the sensor measurements. It has to be observed that the behavior-confidence correspondence by means of fuzzy logic inference is non-linear; linear correspondence would not be adequate in some cases (for example, the confidence deserved by a steady measurement is high; however the confidence deserved by a completely steady measurement is low).

Hence, the inputs to the expert system are not just the sensor measurements but the sensor measurements *plus* the confidence values. The advantages of the proposed architecture are:

- At knowledge acquisition meetings the following statement was frequently obtained: "if this thing occurs then, or there exists a fault in the sensors, or this problem is in progress and ...". With the new validation module the diagnosis problem is separated from the sensor validation problem.
- The system deals with these two problems in parallel, making a significant improvement in the speed and maintainability of the system.
- The confidence values are stored in the blackboard. Thus they can be used by any system module, similarly to the scheme used by human experts: they first check for the reliability of the measurements and then use them for subsequent reasoning. For example the simulator module may calculate a variable by choosing among different formulas, depending on the confidence values of the input variables.

Several other approaches to sensor fusion have been proposed in the literature (e.g. [2], [3], [6], [7], [8]). The following common conclusions can be found in these references:

- The *history* of each measurement from the sensors has to be stored and analyzed to characterize the sensor behavior and process environment characteristics.
- The use of *redundancy* is fundamental for fusing measurements. For instance, we can compare
  - multiple sensors measuring the same physical variable.
  - several sensors measuring simultaneously different physical variables that are numerically related to each other
  - a sensor measuring a specific variable and an estimation of the value of that variable.

- Process experts and plant operators have a lot of experience in recognizing the *behavior* characteristics of specific sensors. So, the distinction between normal and abnormal behavior in each sensor have to be based on that experience. For example, if we are measuring the temperature of a bulk of material, the measurements must change slowly with time: quick changes are not expected. On the other hand, if the sensor is measuring a pressure or flow of material through a valve, steady values are suspicious: pressure and flow measurements are typically noisy.

It has to be remarked also that two different sources of *uncertainty* arise from any measurement from the sensors: the *imprecision* induced by systematic and random errors in the sensor measurements because of technical limitations, and the occurrence of *faults* in the sensors because of the process environment characteristics.

In the rest of the paper we describe the fuzzy logic based techniques used to help with sensor fusion, taking into account the conclusions cited above.

2. Fuzzy Logic Techniques for Sensor Fusion

In this paper sensor fusion is proposed as a complementary module for the MIP expert system by means of monitoring the incoming data and assigning confidence values to each sensor measurement. For calculating these values the module considers the *individual* behavior of each sensor and the *collective* behavior of groups of related sensors.

2.1 Individual behavior

The history of the measurements is processed to analyze the specific individual characteristics; for example the work range of the sensor, the differences between two consecutive values, the second order differences, an estimation of the current direction of the medium term change, etc.

These characteristics reflect both the sensor and process behavior. From the history, we construct a histogram for each of the characteristics and from the histogram a possibility distribution is defined ([4], [5], see figure 1). Next, the current values of the possibility distributions for each of the characteristics are evaluated. From these estimations an *individual confidence value* for the sensor is calculated by a t-norm (for instance, the minimum). The obtained value is interpreted as the *normality* of the sensor individual behavior.

At this point, we need to determine how much time is necessary to consider for determining the possibility distributions. The expert system stores 96 hours logs of each measurement. Two possibility distributions obtained from two consecutive 96 hours logs are shown in figure 2; the distributions are very close. If we take several logs of each month, the distribution evolution for longer periods of time can be observed (see figure 3).

Our proposed solution is to take a 96 hours log to get an initial distribution and then calculate a new possibility distribution every 3 hours, in parallel with the assignment of confidence values.

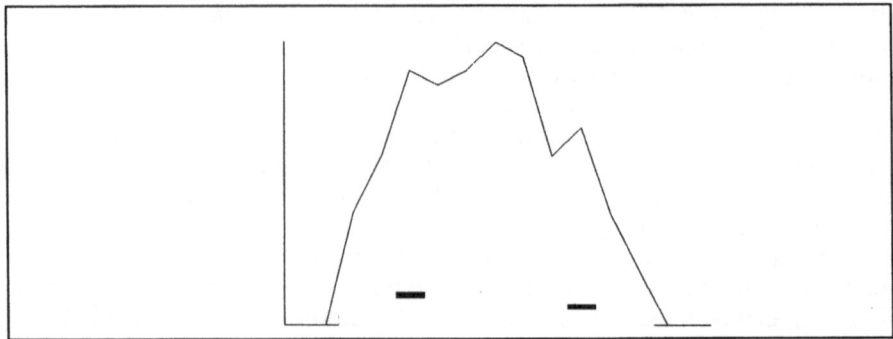

Fig. 1: Possibility distribution obtained from a histogram.

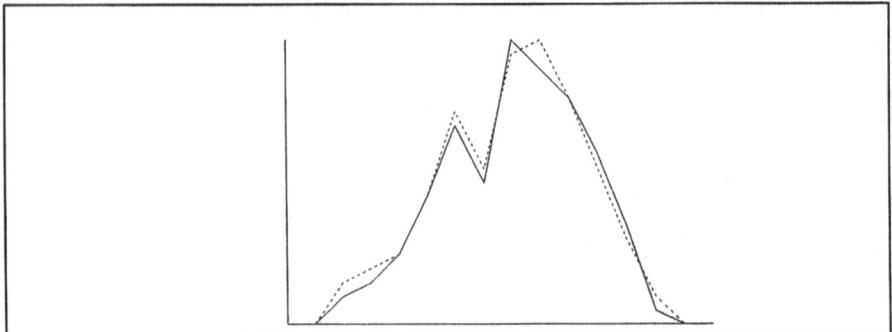

Fig. 2: Two consecutive 96 hours possibility distributions.

These two distributions are combined every 3 hours using the methods proposed in [4] (pp. 146-147). So, the resultant possibility distribution is continuously evolving from an initial state to a state closer to current sensor behavior.

2.2 Collective behavior

After considering the individual behavior, we can analyze the collective behavior of groups of related sensors. Two different cases have to be distinguished: several sensors measuring independently the same physical variable or several sensors measuring simultaneously several numerically related physical variables. In both cases, the expert knowledge about the collective behavior of the sensors is expressed in terms of fuzzy rules which monitor on-line this behavior. If we are measuring the same physical variable, the rules determine the grade of coincidence in the values and in the evolution of the values. On the other hand, if we are measuring different related variables, the rules reflect the agreement of the variables evolution with their relationships: several *collective confidence values* are ascertained from the rules by means of fuzzy logic inference. In the first case, pairwise confidence values are obtained to each measurement. In the second, one collective confidence value is assigned to each sensor.

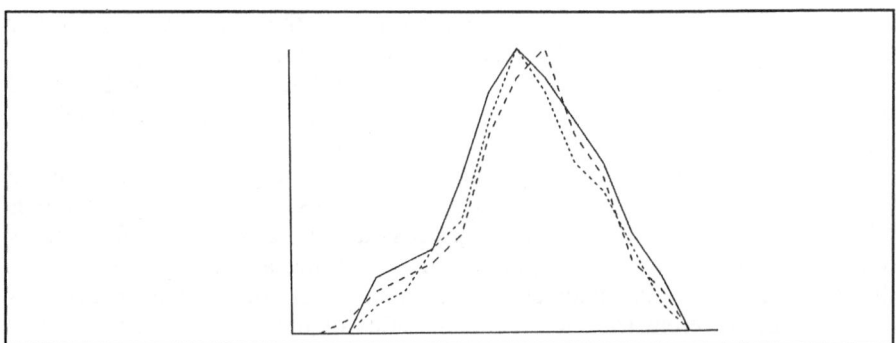

Fig. 3: Three possibility distributions from three different months.

At the end of the proposed process, we fuse both the individual confidence value and the several collective confidence values into a final confidence value by means of a suitable function. For instance, this can be done by

$$\sqrt{w_1\,\alpha_1^2 + w_2\,\alpha_2^2 + ... + w_n\,\alpha_n^2}$$

where α_i are the confidence values obtained, and w_i are the weights of each confidence value in the final fusion ($0 < w_i < 1$, $w_1 + ... + w_n = 1$).

The weights are accordingly determined with the expert knowledge about the characteristics of the sensors (e. g. precision, sensitivity, reliability expected). The closer to 1 the confidence values are, the closer to 1 the final confidence value. In this way it is very easy to tune the impact of each confidence value in the final fusion and the function presents in practice smoother changes than a linearly weighted average.

The final confidence value is associated to the measurement and stored in the blackboard. Hence, any other system module can use it for posterior reasoning.

3. Case Study

We have selected two particular cases to show the procedure described in the preceding sections. These examples have been selected because they can be easily generalized to any similar groups of sensor measurements.

3.1 Case Study 1: Oxygen Excess

The oxygen excess produced in the output of the petrochemical plant is a critical variable for supervisory control: there are four sensors in total measuring this variable in the plant. Two of the measurements (A and B) come from field sensors with different sensitivity and reliability factors.
The other two (C y D) come from an automatic analysis from a mass spectrometer. Besides, after an experiment realized in the pilot plant (a miniaturized reproduction of the entire plant) the expert found out an accurate

calculation of the oxygen excess in the reactor of the plant based on other sensor measurements such as the reactor control temperature, the pressure on the top of the reactor, and the molar relations air/propylen and ammonia/propylen. In this way, we have an additional estimation (E) to compare with the direct measurements.

However, there are some problems in the comparisons. First, these different measurements are not synchronized because the mass spectrometer needs approximately seven minutes to process a sample, and the field sensors take approximately fourteen minutes for analyzing the same sample. Second, it is better to compare the changes in the values than the values themselves because of the different technical characteristics of the sensors: all the values are similar but they are not exactly the same, while the changes in the values should be almost equal for all sensors.

Therefore, four confidence values are calculated: three pairwise confidence values and one individual value. This one is obtained from the different possibility distributions, as explained in the previous section. The pairwise confidence values are determined by comparing the values and the change in the values of the sensors and deciding their evolution coincidence rate, by means of a set of fuzzy rules such as:

- If sensor A is STEADY and sensor B INCREASED-A-LOT, then pairwise confidence AB is LOW and pairwise confidence BA is MEDIUM.
- If sensor A is STEADY and sensor D is STEADY, then pairwise confidence AD is HIGH and pairwise confidence DA is HIGH.
- If sensor C INCREASED and sensor D DECREASED, then pairwise confidence CD is LOW and pairwise confidence DC is LOW.
- If sensor B INCREASED and sensor C INCREASED and sensor E INCREASED-A-LOT, then pairwise confidence BC is MEDIUM and pairwise confidence CB is MEDIUM.

The rules and the impact of each confidence value in the final result are designed with the available knowledge from the expert about the specific characteristics and behaviors of the sensor measurements.

The final fusion of all the confidence values for each sensor is performed by the function shown in the previous section.

3.2 Case Study 2: Ammonia Evaporator

In the ammonia evaporator of the petrochemical plant considered in the MIP project there are three sensors: temperature, pressure and ammonia level. Given the physical and chemical structure of the evaporator, these three sensors can be considered as three different measurements of the state of the evaporator; so, it is important to know something about the reliability of these measurements. They have to be related as follows:

- If the level decreases a lot, temperature has to increase a lot and pressure has to increase (the opposite is not true).
- Changes in temperature and pressure have to occur in the same direction.
- Temperature has to change at least in the second decimal figure.

We can use in addition the ammonia flow measurement, the state of the valves controlling the ammonia flow, the ammonia molar relation estimate into the evaporator, etc. We compare the evolution of these auxiliary data with the evolution of the three main measurements to ascertain their confidence level. Both the former relationships and the comparisons are translated into fuzzy rules as follows:

- If LEVEL DECREASED-A-LOT and TEMPERATURE INCREASED-A-LOT and PRESSURE INCREASED, then collective confidence LEVEL is HIGH, collective confidence TEMPERATURE is HIGH, and collective confidence PRESSURE is HIGH.
- If TEMPERATURE DECREASED and PRESSURE DECREASED, then collective confidence TEMPERATURE is HIGH and collective confidence PRESSURE is HIGH.
- If ammonia flow is STEADY, the valve is STEADY, and PRESSURE DECREASED-A-LOT, then collective confidence PRESSURE is LOW.

Finally, using the possibility distributions defined for each behavior characteristics and the expert knowledge expressed in fuzzy rules, several confidence values are calculated. The final confidence value is obtained by the method used in the previous example.

4. Conclusions

An approach to fuzzy logic sensor data fusion has been presented in this paper. Aspects that have contributed to its success are the use of the history of the measurements, the information redundancy, the behavior analysis and the easy tuning of the final module.

The approach may be generalized readily to any other groups of sensor measurements. However, there are some problems to solve:

- What happens to the possibility distributions if the history data is defective? Is it necessary to filter defective data?
- It seems reasonable to use also utility values: if some variable is critical, how should this impact in the calculation of the confidence values?
- When is the rule set complete?

Ongoing research is addressing the solution of these problems.

The described procedure for sensor fusion has been implemented as a new module for the MIP expert system. In a first stage the fuzzy logic module was installed in the off-line version of the: MIP system.

The results obtained were quite satisfactory and the module is currently in production in the plant. In this phase other sets of measurements have been added to the module.

The methods developed in this project will be applied in the HINT project, an European ESPRIT project that will combine different Artificial Intelligence techniques (model-based reasoning, neural networks, fuzzy logic, expert systems, etc.) in an integrated architecture with the goal of improving industrial process control operation.

References

1. Alamán X. et al., "Knowledge-Based Systems for Real-Time Process Control: The MIP Project". 1992 IFAC/IFIP/IMACS International Symposium on Artificial Intelligence in Real-Time Control, Delft, The Netherlands, June 1992.
2. Alexander et al., "An Architecture for Sensor Fusion in Intelligent Process Monitoring", Computers Ind. Eng. Vol. 16, No. 2, pp. 307-311, 1989.
3. Chendrasekaran B. and Punch W. F., "Data Validation During Diagnosis, A Step Beyond Traditional Sensor Validation", AAAI 1987, pp. 778-782, 1987.
4. Dubois D. and Prade H., "Possibility Theory", Plenum Press, New York, 1988.
5. Dubois D. and Prade H., "Fuzzy Sets, Probability and Measurement", European Journal of Operational Research Vol. 40, pp. 135-154, 1989.
6. Fox M.S. et al., "Techniques for Sensor-Based Diagnosis", IICAI 83, pp. 158-163, 1983.
7. Leinweber D. and Gidwani K., "Real-Time Expert System Development Techniques and Applications", Western Conference on K.B. Engineering and E.S., pp. 69-77, 1986.
8. Rowan D. A., "On-Line Expert Systems in Process Industries", AI EXPERT, Miller Ferman Publications, August 1989.

Task-Directed Sensor-Planning

Gerhard Grunwald
DLR German Aerospace Research Establishment
Institute of Robotics and System-Dynamics
Münchenerstr. 20
D-8031 Oberpfaffenhofen

Gregory D. Hager
Department of Computer Science
P.O. Box 2158 Yale Station
Yale University
New Haven, CT 06520

Abstract

Recently, there has been increased emphasis on employing reactive actions in robot task planning. The principle reasons for this change are to increase the robustness of robot actions by making them sensor controlled, and to accommodate dynamic, unpredictable environments. However, in many cases, supporting reactive mechanisms requires choosing sensor inputs for the reactive procedure. This paper addresses the issue of planning the sensing and fusion required to carry out a reactive robot program. A preliminary framework for planning and fusion is presented, and sensor planning is illustrated for the problem of replacing a mechanically attached plug in a space environment.

1   Introduction

In recent years, there has been a substantial shift in the philosophy of connecting perception to action. Formerly, it was common to assume that robot planning and execution had a complete, exact world model available. This model was either an artificial geometric model calibrated offline with the robot coordinate system, or was to be recovered online from sensor information. However, the former approach depends heavily on global calibration and consequently is extremely non-robust. The latter approach assumes that the robot carries sufficient sensing and computational capabilities to perform reconstruction in all circumstances at a refresh rate sufficient to satisfy all world dynamics. For many applications, particularly in unstructured, dynamic environments, both approaches are patently impractical.

Recently, the "purposive" or "reactive" philosophy of connecting sensing and perception has been gaining wider acceptance [1, 2, 3, 14]. In reactive robotic systems, robot action is continuously and directly conditioned on sensor inputs. There is no intervening reconstruction. Direct sensor feedback makes it possible to specify actions in *relative* coordinates rather than in *absolute* coordinates as in a world-model-based system. Consequently, model discrepancies and/or environmental variations are instantly and automatically adapted to, making the system much more robust and reliable.

There have been many proposals for reactive programming and planning systems [4, 12]. While reactive task planners have addressed the problem of decomposing tasks into reactive actions, few systems have considered the problem of choosing the sensor information needed to support those actions. In this context, sensor planning needs to consider two main issues: what information is needed to carry out an action, and what collection of sensor inputs can supply that information. The planner must ensure that sufficient information is available to maintain stability and to provide for robustness against sensor error, but must guard against overwhelming the system with large quantities of information that cannot be processed at servo rates.

In this paper we describe preliminary work on a sensor planning system for use in structured environments, and illustrate those ideas on a problem drawn from the domain of space teleoperation. We focus on geometric operations that involve recognition and manipulation of known objects in a structured environment. The major sources of information are vision, range, and contact sensing.

2 Planning System Structure

We view sensor planning as a generic operation that could support any number of client applications including a task-specific robot program, an assembly planner, or a teleoperated robot system. We assume that each client simply generates a sequence of robot actions, some of which require sensor support. The sensor planner takes this "program" and attempts to construct a sequence of sensing operations that supply the information needed to carry out the program. If the sensor planner cannot create a plan that fulfills the information requirements of every action, it returns a failure message to the client along with an indication of why the action cannot be carried out. It is then

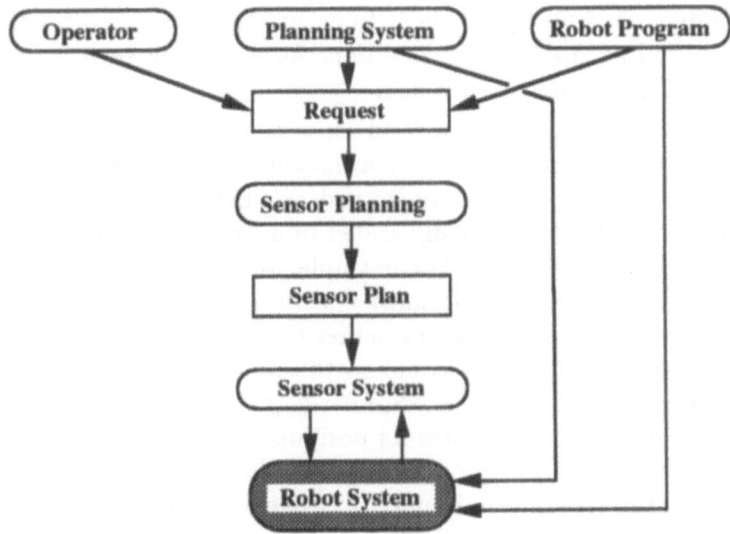

Fig. 1: The global structure of the sensor planning system

up to the client to revise the action sequence and resubmit it to the sensor planner.

2.1 Action Types

As noted above, a robot program consists of a series of actions. Examples of these actions include:

- *grasp (OBJ)*: grasp *OBJ* at predefined grasping points ;

- *release ()*: release the currently held object;

- *disconnect (OBJ)*: execute a specialized procedure for disconnecting *OBJ* from a fitting;

- *recognize (OBJ)*: find *OBJ* in a scene;

- *pose (OBJ)*: determine the position and orientation of *OBJ*;

- *approach (OBJ)*: move the gripper to a position relative defined to *OBJ*;

- *follow (OBJ)*: track the motions of *OBJ*.

One class of actions consists of procedures like *grasp, release,* and *disconnect* that, while they may rely on some type of sensing, do not require any sensor planning. It is up to the client to ensure that the robot program has been constructed so that the preconditions for executing such actions have been achieved before they are invoked.

The remaining actions require sensor support. The type of support needed is classified as a *static* sensor task or a *dynamic* sensor task. Static sensor tasks perform one-time measurements and calculations. These actions are purely informatory, and are usually added to a program to instantiate variables needed by later actions. For example, *recognize(OBJ)* addresses the problem of matching sensor information with a model, *OBJ*. This information can be used to relativize an action to a particular object, to verify expected initial conditions, or to aid in the search for specific object features needed to control an action. Similarly, *pose(OBJ)* assumes that *OBJ* has been recognized and computes its position. There are a wide variety of possibilities for recognizing objects and computing pose. In fact, some types of geometric-based recognition implicitly or explicitly compute pose, making *pose* redundant [5, 11]. However, the client has no *a priori* information on how recognition will be carried out. Consequently, it is the job of the sensor planner to make sure that recognition and pose estimation are matched in the appropriate fashion.

Dynamic sensor tasks support control of *performatory actions*: actions that lead to actual motion of the robot. For example, approaching an object requires tracking lines or patterns that can control the attitude and relative position of the robot with respect to the goal position. The job of the sensor planner is to choose the type of sensing and the sensor patterns to be used in the feedback loop. It must ensure that the entire action, from initial state to final state can be carried out using some combination of sensing and servoing. The connection between actions proposed by the task-level planner and the operations planned at the sensor level takes two forms. For informatory actions, the action generated by the task planner corresponds directly to a task to be planned by the sensor planner. In some cases, the task-level planner is allowed to specify accuracy requirements or other conditions that must be met. For example, a pose estimate may need to meet accuracy requirements in order to support a particular fine motion strategy. Performatory actions that require sensing support are "decomposed" into sensing tasks by the sensor planner. This decomposition is governed by the contextual constraints surrounding the action. For example, approaching an object requires recognition, pose estimation, and tracking to support control. The sensor planner also has information on the type of control that is taking place so that it can choose the features to be tracked based on control stability criteria. We discuss these issues in greater depth in the next section.

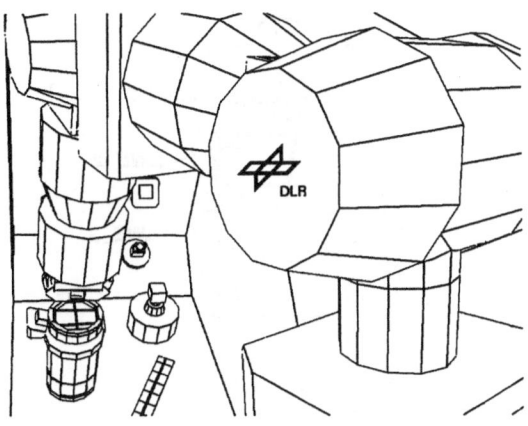

Fig. 2: The ROTEX workcell

2.2 Experimental Planning Environment and Example.

Our experimental environment consists of the ROTEX (RObot Technology EXperiment) robot, sensors and workcell. ROTEX is a German space telerobotic experiment that will fly with the next German space lab mission (D2) in 1993. It consists of a six-axis robot equipped with a sensorized gripper and an external fixed pair of video cameras providing a stereo image of the robot workspace. The gripper is provided with a number of sensors: two six axis force-torque wrist sensors, two tactile arrays, grasping force control, an array of 9 laser-range finders and a tiny pair of stereo cameras. ROTEX is designed to operate autonomously, to be teleoperated by astronauts, or to be teleoperated from a ground station. A more detailed description of the ROTEX environment can be found in [9] or [10]. Although the ROTEX environment was primarily designed for telemanipulation, there are additional interfaces in the laboratory version of ROTEX that support other modes of operation. We note that, for the moment, the configuration used for sensor planning experiments only uses a single hand-camera and a single global-camera: no stereo information is available.

As discussed in a previous paper [7], sensor planning in a teleoperated system reduces the likelihood of execution error by maximizing the autonomous, sensor-based execution of tasks. An operator formulates a task by entering commands with a tele-command device. The specified task can either be executed autonomously by the remote system or executed as a semi-autonomous cooperation between the sensor system and the operator. The man-machine cooperation will be useful if the sensor system cannot constrain all directions of motion. For example the operator may command the gripper to approach an object in a certain configuration. If the sensing system cannot find sensor

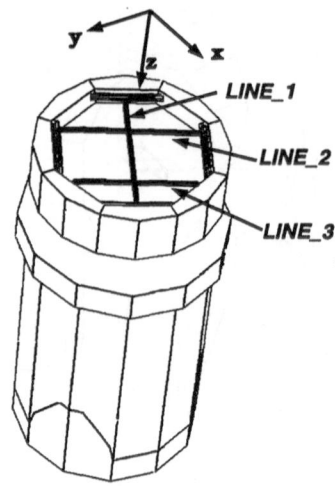

Fig. 3: The Qrbit Replaceable Unit

inputs to control the orientation of the gripper, it will turn those degrees of freedom over to the operator while maintaining the proper values for gripper position through sensor feedback.

Our sensor planning framework will be illustrated using one of the D2 mission experiments: the task of finding, grasping and removing the *ORU* (Qrbit Replaceable Unit), an electro mechanical plug (See figure 3). The *ORU* is removed by expanding a grip on the top, executing the proper twisting motion, and then pulling away from the attachment surface. We will consider planning for the recognition and sensor-guided approach to the *ORU*

We note that experiments with the system in teleoperated mode have shown that this task is extremely difficult to carry out with no sensor feedback. The tolerances on the hand position and orientation are very small, and the sensor information available to the human operator is often difficult to interpret. The system permits the operator to use the four laser distance sensors pointing out of the fingertips to control gripper orientation near the *ORU*. This simplifies the task greatly. We have recently developed a vision-based controller that can control gripper position parallel to the surface of the *ORU*. Combining both range and vision leaves the operator the simple task of getting the robot near the *ORU*, turning on sensor-based control, and then controlling the z axis motion until the gripping position is reached. Our goal is build a scnsor planning system that can automatically invoke these procedures without explicit operator intervention.

3 Planning Sensor Tasks

As noted in the previous section, requests to the sensor planner are either basic sensor requests or higher level actions that are decomposed by the sensor planner into a sequence of more basic sensing actions. For example, the high level action *approach(ORU,final_pos)* is decomposed into the following operations: *approach(ORU,final_pos)* :=[*recognize(·); pose(·); tracking(·)*]. The first two operations are static, informatory actions, while the latter is used to support performance of a dynamic closed-loop motion. The sensor planner constructs a plan for these actions using information available on the *ORU*, the available sensors, and final position criteria supplied in the request.

Information on objects and sensors comes from object models, sensor models, and sensor simulations. The object models are from the CATIA CAD modeler which provides us with a boundary representation from which physical edges and planar surfaces can be easily extracted. In addition, we separately represent artificial markings such as the guide lines on the ORU. Using this information, we have implemented and tested simulations for the distance sensors, a laser range sensor, and a 6 dof force/torque sensor. Further we have also implemented a primitive camera simulation which is currently being extended to deal with more sophisticated features and imaging models.

3.1 Logical Sensors

The basic representation in the system is a structure similar to the notion of Logical Sensors [8]. The basic form of the structure is:

Logical Sensor	<name>
Input_List	{<Logical Sensors>}
Data_Proc	{<return> = <data_proc(...)>}
Type	**static	dynamic**
Controller	{<ctrl>}
Valid	{<cond>}
Characteristic	<pose>; <fixed|movable>; <field of view>; <...>

At planning time, these logical sensors are instantiated and organized into a network structure based on various choices of input sources from the *Input_List* and data processing procedures from the *Data_Proc* list. Each input process has a return value which is used as input to a data processing algorithm. The *Type* slot indicates whether this sensor provides information just once, or over an interval of time. The *Valid* slot represents constraints that must be satisfied by the input to the chosen data procedures in order for them to function correctly and reliably. The *Controller* must check the sensor data with respect to *Valid* and, in case of a dynamic operation, coordinate and synchronize the instantiated data processes. If the Logical Sensor represents a physical sensor it may also control the device. Physical sensors use the *Characteristic* slot to represent information like the position, the field of view, work-criterions and very important if its a fixed sensor or its mounted on a movable device.

3.2 Planning for a Static Action

The basic concepts of planning for a static actions can be illustrated with *recognize(ORU)*. A description of the logical sensor *recognize* is:

Logical Sensor	recognize
Input_List	line, region, feature
Data_Proc	matched = match_line (measured_val, model)
	matched = match_region (measured_val, model)
	matched = match_feature (measured_val, model)
Type	static
Controller	valid, quality
Valid	matched
Quality	$[matched - \varepsilon, matched + \varepsilon]$

The *Input_List* describes all logical sensors which could contribute to the recognition process. Likewise, there are several means for recognizing objects given in the *Data_Proc* list. Finally, we will assume the recognition can always be attempted, hence no initial conditions are specified.

In the ROTEX environment, there is a great deal of prior information about the position and appearance of objects. In particular we have information on artificial markings that uniquely identify objects. These marks are already available in space telerobotics because they are used as visual support for the teleoperator. This feature information can be used by the sensor planning system to further instantiate the logical sensor *recognize*. Specifically, the sensor simulation system may decompose *line* into a line detector running on the hand camera, or on the external camera. For each case, it can use expected object position to check features for observability using techniques similar to [13], and thereby compute the efficancy of each camera for the task of recognizing the *ORU*. These evaluations can be used to select the best sensor for the expected state of affairs. Finally, simulation values of the *ORU* features can be used to reduce the area considered by the line detection and also to reduce the combinatorics of model matching.

Putting all of this together leads to the following schematic sensor plan for *recognize (ORU):*

Sensor:	*Hand Camera*
Operation Type:	*static*
Algorithms:	*extract_lines (image)*
	match_line (measured_val, model)
Cooperation:	*none*
Start Condition:	*none*
Goal Condition:	*ORU found*

The sensor plan is designated to work with information from the physical *hand camera* sensor. It further specifies all of the initialization and parameter information needed to support autonomous execution of the recognition task. Recognize is a static (one-time) operation; consequently the camera will not be required for a long period of time. The processing requires two algorithms: an edge extraction operation, and a line matching operation. Since the teleoperation environment already has rough position information on the position of the *ORU*, matching is given an expected pose of the object to speed up operation. In this case, recognition is really performing a verification operation. The cooperation parameter is used to define other sensors which could contribute to the recognition process. Here there are none, but an example below will include cooperation. The start and goal condition are defined either by the client or the sensor planner itself to give additional constraints to the algorithms. In this case, the goal condition does not provide any additional information.

3.3   Planning Dynamic Actions

Approach (ORU) seeks to move the manipulator into grasping position using constant sensor feedback to adjust the position and attitude of the gripper. From the planner's perspective, dynamic actions are performing a control operation that is decomposed into: device degrees of freedom that are actively servoed without vision or range sensor feedback, degrees of freedom that are controlled by vision or range sensor feedback, and a termination condition that signals the end of the action. We will henceforth refer to the actively servoed degrees of freedom as *driven*, and the remaining degrees of freedom as *regulated*. In the case of approach, the z direction of motion relative to the tool frame is driven while the other degrees of translational and rotational freedom are regulated. The action stops when the gripper is within 3 millimeters of the *ORU* center, perpendicular one millimeter above the *ORU* surface, and the gripper jaws are parallel with two of the four release pins.

grip configuration 1

grip configuration 2

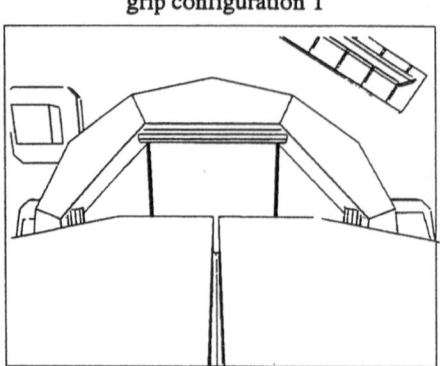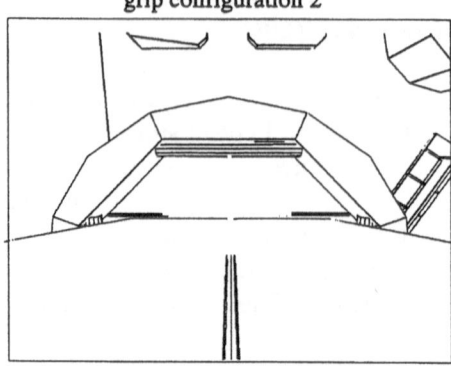

Fig. 4: The two possible gripper configurations for grasping the ORU

An example of a logical sensor for dynamic operations is line tracking:

Logical Sensor	line_tracking
Input_List	line
Data_Proc	line_track (line)
Type	dynamic
Controller	valid
Valid	line_location

The major difference from previous examples is that this is classed as a dynamic sensor, and that line tracking requires an initial line location to start the tracking process. In the dynamic case, the major problem for sensor planning is to ascertain whether or to what extent a device can be controlled along a path from initial to final position using collections of the available logical sensors. If not enough information is available, sensor planning refuses the request and informs the client what type of information is missing. As noted in the introduction, the goal is to find enough information for stable, robust control without overwhelming the feedback algorithm with sensor data.

The first step in the planning operation is to interpret the goal conditions in terms of sensor information by using the sensor simulation system. In the case above, the gripper is simulated positioned parallel and centered one millimeter above the *ORU* surface. If, as in this case, the goal conditions are not unique (there are two possible orientations of the hand that satisfy the conditions), the planner notes every possible configuration. The simulated sensor outputs in each goal configuration constitute a sensor-based interpretation of the goal conditions (see figure 4).

Now, the central problem for the planner is to evaluate stability of controlling the gripper with respect to the regulated degrees of freedom. This evaluation proceeds as follows: for each logical sensor potentially used as feedback for the controller, we assume that we are given sensor model of the form

$$z = H(p, d) + V.$$

A discussion of such models can be found in [6]. The parameters p constitute the regulated parameters, while d denotes driven (or other unregulated) degrees of freedom.

Using this model and a previously calculated goal position, p_0, the sensor simulation generates the nominally expected sensor outputs, z_0. The control problem is to maintain $H(p, d) = z_0$ over an appropriate range of system configurations p and d. In practice, the control problem often more easily and naturally expressed in terms of the minimization of an objective function. Since maintaining an invariant can be expressed as minimizing the objective $O(p, d, z) = \|H(p, d) - z\|^2$, we will henceforth adopt this point of view. Let $\mathcal{S}$ be a set of logical sensors, O_s be the control objective function for a sensor s, and z_s be data from sensor s. At a time instant, the global objective function is:

$$O(p, d, \mathcal{S}) := \sum_{s \in \mathcal{S}} O_s(p, d, z_s)$$

The local stability of a solution for regulated parameters depends on investigating the eigenvalue structure of the Hessian of the objective function at the point of interest. Computing the Hessian of the objective function and substituting values from the sensor model evaluated at the point where stability is to be evaluated leads to an expression of the form:

$$\nabla^2 O(p, d, \mathcal{S}) := \sum_{s \in \mathcal{S}} \nabla^2 O_s(p, d, H_s(p, d)).$$

The vector of eigenvalues can be reduced to a single value in any number of ways. We will compute the product of the eigenvalues which is proportional to the area of the ellipsoid defined by the Hessian matrix. Maximizing this measure by varying the members of $\mathcal{S}$ permits us to choose the collection of inputs that maximize the stability of regulated motions.

Returning to our example, data from the four laser distance sensors constrains translation in the z direction and rotation around the x and y axes. In other words, these sensors can enforce the condition of orthogonality. The other three dofs must be constrained by the hand-camera. To take advantage of relative control the geometric information from the image should rely on both the position of the ORU and that of the gripper. The possible features of the gripper are the horizontal-lines, g_1 and g_2, parallel to the x axis of the image and the small gap between the fingers which appears in the image as the line, g_3, parallel to the y axis. Referring to figure 3, the potentially usable ORU features are the two dark parallel lines l_1 on the top and l_2 on the bottom, and the single perpendicular line l_3 in the middle.

Noting that l_3 is not visible from grip configuration one, we evaluate the Hessian using the laser distance sensors and observations of l_1 and l_2. This yields eigenvalues of:

$$11.9917, \quad 10.204, \quad 3.41027, \quad 0.260319, \quad 8.13368, \quad 0.$$

The final zero indicates that the Hessian is not full rank, and hence some degree of freedom cannot be controlled. In this case, we cannot control translation parallel to l_1 and l_2.

In grip configuration two, all three lines are visible. Evaluating the Hessian in this case yields:

$$23.1335, \quad 10.7357, \quad 8.37067, \quad 2.43937, \quad 1.67418, \quad 0.313255$$

with a product of 2659.6. The system is extremely stable. Suppose that a line is mistracked and lost. Dropping l_2 from the calculation yields:

$$10.7824, \quad 10.4721, \quad 2.33953, \quad 1.52786, \quad 0.211425, \quad 8.$$

with a product of 682.667. In other words, some redundance has been lost by dropping the line, but the system is still controllable. Consequently, configuration 2 is preferable to configuration 1 and will be chosen as the goal configuration to attain.

Further away from the ORU, the range sensors can no longer detect the ORU reliably. Using only visual feedback from l_1, l_2, and l_3 leads to eigenvalues of

$$5.07037, \quad 4., \quad 0.262966, \quad 3.81012 * 10^{-18}, \quad 1.00533 * 10^{-16}, \quad 0.0.$$

From this, it is clear that the control is locally unstable with respect to all 6 degrees of freedom. However, the z axis is irrelevant since it will be driven. Furthermore, regardless of orientation about the x and y axes, moving in the tool z direction will eventually lead to a configuration where the laser sensors can detect the ORU. Hence, there is no need to control x and y orientation when far from the ORU. Fixing those degrees of freedom and neglecting translation in z yields eigenvalues of

$$18.2481, \quad 5.65462, \quad 2.50931, \quad 0.791176, \quad 1.16317, \quad 0.300298$$

with product 71.556. This again indicates that the system is quite controllable. Thus, a good plan is to detect the ORU, move toward it until the visual features are visible, continue moving toward it until the range finders acquire the target, and then to move to the final position under combined visual and range sensor control.

The calculations are collected into two sensor plans necessary for the hand-camera and the distance sensors. In contrast to the previous sensor plan these are defined for dynamic actions. Cooperation must be specified explicitly as both sensors must work together to constrain all degrees of freedom. A schematic sensor plan for the hand-camera is:

Sensor:	*Hand Camera*
Algorithms:	*lines = extract_lines (image)*
	measured = match_line ([l_1,l_2,l_3], lines,[m_1,m_2,m_3])
	line_track(m_1), line_track(m_2), line_track(m_3)
Operation Type:	*dynamic*
Cooperation:	*Distance_Sensors*
Start Condition:	*none*
Goal Condition:	$\|m_1 - g_1\| \leq$ *1.5 mm*
	$\|m_2 - g_2\| \leq 1.5$ *mm*
	$\|m_3 - g_3\| \leq 1.5$ *mm*

The sensor plan for the laser distance sensors is analogously formulated with a cooperation reference to the hand-camera. The only difference is a start condition based on the distance to the *ORU* surface.

We have experimented with vision-based control for the *ORU* approach using an specialized adaptive controller we have developed. Our stability results concur with the analysis. In addition, we note that if the gripper is tilted with respect to the *ORU* surface, the vision-based control is better conditioned than when it is parallel. Consequently, it is in fact possible to weakly control at least one orientation during the *ORU* approach from visual information alone.

4   Discussion

We have outlined the general structure of a sensor planning system, and illustrated the approach for a concrete problem in the ROTEX teleoperation environment. Our implementation of these ideas is still in the early stages, and hence many details have not yet been specified. One major issue to be dealt with is a general way of turning the local stability analysis using Hessian matrices into a global stability analysis. As a first approach, we plan to simulate the control of the device backwards from the goal and/or forward from the initial state. Whenever new information becomes available or old information disappears, the rank of the Hessian at that point is evaluated.

If the transition preserves the rank, then the change is recorded and will be accommodated at runtime. If the rank drops below an acceptable range, then a search is initiated in the area for a collection of sensor inputs that preserves rank. If one is found, this transition point is recorded as an intermediate goal. If none is found, then an error is recorded. This indicates that operator guidance is needed, open-loop control must be used, a programming error, or may have other context-dependent interpretations.

References

1. J. Aloimonos. Purposive and qualitative vision. Proc. Image Understanding Workshop, pages 816 – 828, 1990.

2. R. Brooks. A robust layered control system for a mobile robot. IEEE Journal of Robotics and Automation, 1986.

3. T. Dean and M. Wellman. Planning and Control. Morgan Kaufmann, San Mateo, CA, 1991.

4. J. Firby. An investigation into reactive planning in complex domains. Proceedings of Sixth Conference on Artificial Intelligence, pages 202 – 206, 1987.

5. W. Grimson and T. Lozano-Perez. Search and sensing strategies for recognition and localization of two and three dimensional objects. Third International Symposium of Robotics Research, pages 81 – 88. Gouvieux, France, 1985.

6. G. Hager. Task Directed Sensor Fusion and Planning: A Computational Approach. Kluwer Academic, 1990.

7. G. Hager. Using resource-bounded sensing in telerobotics. 5th International Conference on Advanced Robotics, ICAR, 1991.

8. T. Henderson and E. Shilcrat. Logical sensor systems. Journal of Robotics Systems, 1(2):pages 169–193, 1984.

9. G. Hirzinger. The telerobotic concepts of ROTEX—Germany's first step into space robotics. 39th I.A.F., Bangalore, India, 1988.

10. G. Hirzinger, G. Grunwald, B. Brunner, and H. Heindl. A sensor-based telerobotic system for the space robot experiment rotex. 2. International Symposium on Experimental Robotics. Toulouse, France, 1991.

11. D. Huttenlocher and S. Ullman. Object recognition using alignment. Int. Conf. on Computer Vision, 1987.

12. D. McDermott. Planning reactive behavior: A progress report. In Innovative Approaches to Planning, Scheduling and Control, pages 450 – 458. Allen, J. and Hendler, J. and Tate, A., 1990.

13. K. Tarabanis, R. Tsai, and P. Allen. Automated sensor planning for robotic vision tasks. Proceedings of the IEEE International Conference on Robotics and Automation, pages 76 – 82, 1991.

14. M.J. Tarr and M.J. Black. A computational and evolutionary perspective on the role of representation in vision. Technical report, Technical Report YALEU/DCS/RR-899, Yale Department of Computer Science, 1991.

Incremental Map-making in Indoor Environments

A. Ekman, D. Strömberg
National Defence Research Establishment,
Dept. of Information Technology, Linköping, Sweden

Abstract

This paper presents an algorithm that transforms laser range data from indoor environments, into a bird's eye view of the robot surroundings. The algorithm is intended to be used for semi-autonomous robots, where the operator needs a global view of the environment. Emphasis is put on the important Line Extraction step, where two approaches are mentioned, one of which is a new Polyline Segmentation approach. Input data comes from odometry and a laser range scanner.

1. Introduction

In critical applications, such as bomb disposal, a semi-autonomous robot, monitored by an expert operator, is superior to an autonomous robot.

A problem that has been found during experiments is the problem for an operator to perceive the robot position in the environment if only views from TV-cameras are available.

The Map-making algorithm is intended to reduce this problem by providing a bird's eye view of the robot surroundings for the operator.

The complete algorithm can be divided into two main steps; estimating the position, and maintaining the map database. In figure 1, the three rightmost boxes describe the first step, while the two boxes to the left describe the second step.

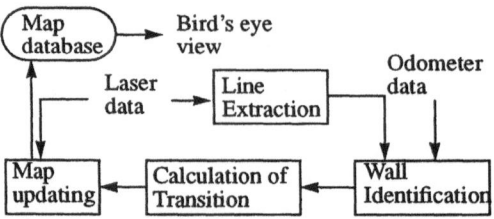

Fig. 1: Map-making algorithm outline

Refer to figure 2 when reading the definitions below.

The state of the robot is defined as:

$$\bar{\xi}^T = \begin{bmatrix} s & \gamma & \tau \end{bmatrix} \tag{1}$$

where s and γ corresponds to position and τ to orientation of the robot vehicle.

The range data tuple is defined as:

$$S = ((r_1, \varphi_1), (r_2, \varphi_2), \ldots, (r_{3200}, \varphi_{3200})) \tag{2}$$

The transition vector is defined as:

$$\Delta\bar{\xi}^T = \begin{bmatrix} \Delta s & \Delta\gamma & \Delta\tau \end{bmatrix} \tag{3}$$

and the odometer measurement of the transition is defined as:

$$\Delta\bar{\xi}^T_{odo} = \begin{bmatrix} \Delta s_{odo} & \Delta\gamma_{odo} & \Delta\tau_{odo} \end{bmatrix} \tag{4}$$

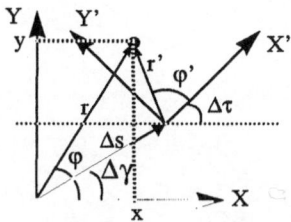

Fig. 2: Mapping of measurement points between two coordinate systems

2. Line Extraction

The primitives to be used are the walls in the indoor environment. To be able to identify a wall in a new measurement as a wall in a previous one, it is necessary to perform a mapping based on the state transition ($\Delta\bar{\xi}$), which is indicated in figure 3.

Equations (5) and (6) summarizes the parameter dependencies that are found in figure 3.

$$p = p' + \Delta s \cdot \cos(\theta' + \Delta\tau - \Delta\gamma) \tag{5}$$

$$\theta = \theta' + \Delta\tau \tag{6}$$

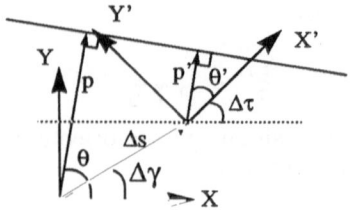

Fig. 3: Coordinates of a wall

The description of a wall in two different coordinate systems, corresponding to two different scan positions, is presented in figure 3.

The appearance of the range data is shown in figure 4.

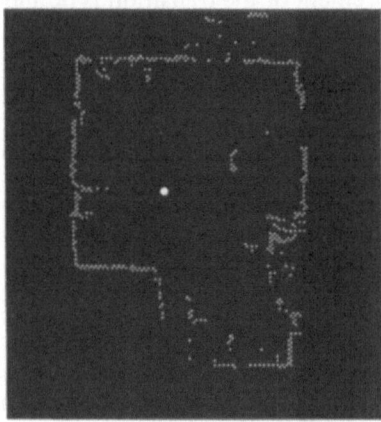

Fig. 4: Typical input data from range sensor

In figure 4, the number of measurement points (m) is 3200. The position of the robot when the scan was taken, is marked with a white dot.

Two different line extraction methods are suggested. The Hough transform was the first approach, while the Polyline Segmentation Algorithm was developed as a faster, and in many cases more precise, line extraction method.

2.1 Hough Transform

Hough transform (HT) is widely used for line detection. It is tolerant to noise, but computational heavy. In the case of range data, though, the HT can be computed rather fast, provided moderate resolution is used in the HT calculation.

Below follows a very brief description of the developed HT algorithm. A more thorough description is presented in [5].

An illustration in the spatial domain of the principle of a HT is shown in figure .

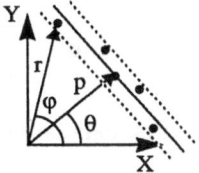

Fig. 5: Illustration of a HT in spatial domain

Notice that, in figure 5, each line (out of a discrete, predetermined set) is tested for how many measurement points it covers. A point is assumed to be covered when it lies inside the area bounded by the two broken lines, as visualized in figure 5.

It is fruitful to transform the contents of figure to the Hough domain:

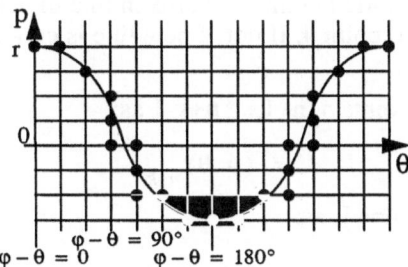

Fig. 6: Spatial points and lines visualized in Hough domain

The cosine curve in figure 6 maps to the measurement point that lies on the tip of the arrow in figure , while the leftmost point in figure 6 corresponds to the solid line in figure . In figure 6, the number of points in vertical direction is depending on the derivative of the cosine curve.

To get an understanding for the Hough transform calculation, it is useful to think of m cosine curves (one for each measurement point), each one having a unique phase angle (φ). For each curve, an iteration is performed in angular direction (θ), and for each angle, a certain number of points are incremented (see figure 6).

The expression below states the same in a mathematical way (assuming $\Delta\theta = 1/a$):

$$p_j = r_i \cdot \cos(\varphi_i - j \cdot \Delta\theta) \qquad \left(\begin{array}{l} i = 1, 2, ..., m \\ j = 1, 2, ..., a \end{array} \right. \qquad (7)$$

2.2 Polyline Segmentation Algorithm (PSA)

The principle of PSA is illustrated in figure 7 and figure 8. DIST corresponds to the distortion, based on an Eigenvector Fit Criterion defined below. Refer to figure 11 for an illustration of the parameters involved.

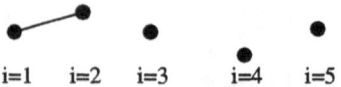

i=1   i=2   i=3   i=4   i=5

Fig. 7: The two first data points are connected

Fig. 8: The line is extended to include a third point

The line in figure 8 is extended to include a third point in a way that minimizes DIST (solid line). Notice the "floating" end-points of the line, reducing the impact if starting in a heavily distorted point. The broken line indicates the position of the line if it is possible to add point 4. If not, a new line is created, starting from point 4.

Define line L, covering points n, n+1,..., n+k-1 as:

$$L = (p, \theta)_{n, k} \tag{8}$$

Define the distortion for line L as:

$$DIST_{n, k} = \frac{1}{k-1} \sum_{i = n}^{n+k-1} \Delta_i^2 \tag{9}$$

where Δ_i is defined as (see figure):

$$\Delta_i = r_i \cdot \cos(\varphi_i - \theta) - p \tag{10}$$

The polyline segmentation algorithm can then be stated as (assuming m data points):

0. n=1

1. k=2

2. Create line between point n and n+1

3. Calculate line $(p, \theta)_{n, k+1}$ minimizing $DIST_{n, k+1}$

4. IF $DIST_{n, k+1} < THRESHOLD$ AND n+k<m:

      k=k+1, $L = (p, \theta)_{n, k}$, GOTO 3

  ELSE

   IF n+k < m-1

     n=n+k, GOTO 1

   ELSE

     HALT

The only non-trivial step in the algorithm is step 3., which is closer investigated in Appendix.

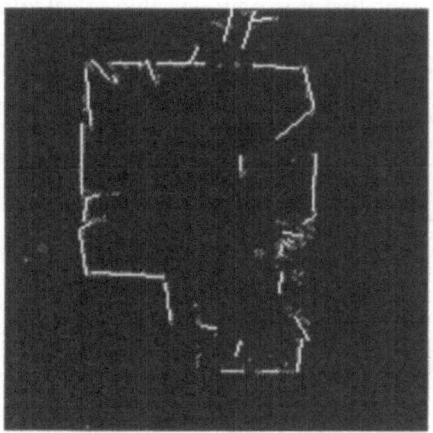 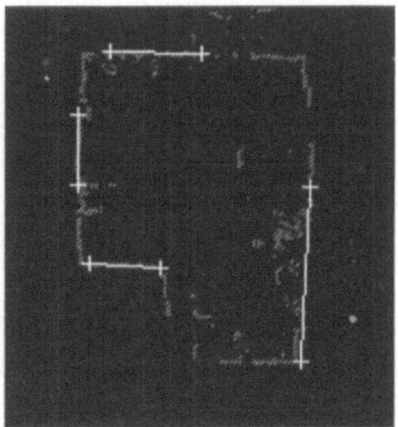

Fig. 9: Application of PSA followed by post-processing

In the left part of figure 9, line segments detected by PSA are superimposed the range data from figure 4. In the right part of figure 9, a post-processing step has been performed on the segments detected by PSA. The output of the post-processing step is shown superimposed the range data from figure 4.

The reason for introducing a post-processing step is to make the algorithm less sensitive to disturbances. Disturbances in the shape of obstacles, occluding parts of a wall are in many cases handled in a proper way, as can be seen in the lower right corner of the right part of figure 9.

3. Wall Identification

This part of the algorithm performs an interpretation from line segments to walls. These walls are compared to the walls extracted in previous scans, to generate a one-to-one mapping between the set of walls. Since walls may become occluded or new walls may become visible, it is necessary to introduce a notation for the state of a wall:

Each wall is either New, Old or Active.

A wall that becomes visible in the current scan is labeled New, while a previously visible wall that has become occluded in this scan is labeled Old. All walls that are visible in both current and previous scans are labelled Active.

The walls with "Active"-labels can be divided into pairs, and are used to calculate the transition in the next step.

An algorithm for conducting the wall identification is given in [5].

4. Calculation of Transition

By inspecting equation (5) and (6) it is obvious that it is now possible to calculate the transition, since (p', θ') corresponds to the wall expressed in the new coordinates and (p, θ) corresponds to the wall expressed in the old coordinates. Each wall pair offers an estimate of $\Delta\bar{\xi}$ calculated through the equations below:

$$\Delta\tau = \theta - \theta' \tag{11}$$

$$\Delta\gamma = -\operatorname{atan}\left(\frac{d_i\cos\theta_j - d_j\cos\theta_i}{d_i\sin\theta_j - d_j\sin\theta_i}\right) + k \cdot \pi \tag{12}$$

where $d_k = p_k - p_k'$.

The range Δs is calculated by inserting the expressions for Δτ and Δγ in equation (5):

$$\Delta s = (p - p') / \cos(\theta + \Delta\gamma) \tag{13}$$

Since Δγ has two solutions, so has also Δs. However, only one solution gives a positive value of Δs.

The different estimates can be combined in a number of ways, for instance a Kalman filter can be used.

5. Map Updating

When the estimate of $\Delta\bar{\xi}$, has been calculated, the local Certainty Grid is created where the modified version of Bayes' theorem given in [6]. Thereafter the global map database can be updated using the relations in figure 2 to transform between the coordinate systems.

The data representation is similar to the one described in [6], with a few changes. One of the most significant is the use of a polar coordinates in the local certainty grid. This conforms to the radial properties in the laser sensor and yields fast updating of the local grid, since the predicate "Cell i lies behind cell j" will be easily verified.

When the local certainty grid has been created, it is mapped to the global grid, which is rectangular. There is no guarantee to have a one-to-one mapping, but this is the prize to pay for higher speed.

6. Presenting Bird's Eye View

The updated, global certainty grid contains the bird's eye view that will be displayed for the operator. Different presentation approaches have been examined; either to output the certainty grid as it is, with the value of each cell to map to a unique magnitude on the display, or to first perform a thresholding to focus on the areas with highest probabilities of being occupied.

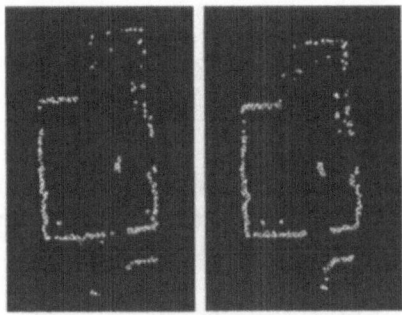

Fig. 10: Two consecutive Certainty grids

The left part of figure 10 corresponds to the Certainty grid created from a first laser scan. The right part of figure 10 corresponds to the Certainty grid created by two consecutive laser scans. Each cell can be related to a colour corresponding to its probability of being occupied. The size of the certainty grids has been 200*200.

7. Conclusions

When counting the number of iterations needed to segment the range data, PSA is very close to optimal; for each measurement point at most two iterations is necessary (see Appendix). In addition, the recursive form of PSA enables a fast execution, independently of the number of points already covered by a line (i e there is no need to recalculate all sums for each new point added).

Though the paper presents a complete Map-making algorithm, it is by no mean argued that this is the best Map-making algorithm, but rather an example of how to make all involved components work together.

Alternative approaches are given in [1], [2] and [3].

8. Appendix

The Appendix contains the derivation of the Polyline Segmentation Algorithm along with a recursive updating scheme.

8.1 Derivation of PSA

Below follows the derivation of the equations needed to perform the calculation of the coordinates of a line (p, θ). The criterion used is Eigenvector fit, as described in [4] (see figure 11).

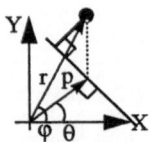

Fig. 11: MSE criterion (broken), and Eigenvector fit criterion (solid bold) for a point (r,φ) according to a line (p,θ)

Δ_i can be rewritten as:

$$\Delta_i = r_i \cdot (\cos\varphi_i \cdot \cos\theta + \sin\varphi_i \cdot \sin\theta) - p \qquad (14)$$

In this equation, $\cos\varphi_i$ and $\sin\varphi_i$ can be calculated in advance and put in a table of size 3200*2. Let c_i denote $\cos(\varphi_i)$ and s_i denote $\sin(\varphi_i)$, where $\varphi_i = i*2\pi/3200$.

It is now possible to express the distortion in the following manner:

$$DIST_{n,k} = \frac{1}{k-1}\left(k \cdot p^2 + \cos^2\theta \cdot \sum_{i=n}^{n+k-1} (r_i c_i)^2 + \right.$$

$$\sin^2\theta \cdot \sum_{i=n}^{n+k-1} (r_i s_i)^2 + 2 \cdot \sin\theta \cdot \cos\theta \cdot \sum_{i=n}^{n+k-1} r_i^2 c_i s_i - \qquad (15)$$

$$\left. 2p\left(\cos\theta \cdot \sum_{i=n}^{n+k-1} r_i c_i + \sin\theta \cdot \sum_{i=n}^{n+k-1} r_i s_i \right) \right)$$

Name the sums A, B, C, D and E respectively:

$$DIST_{n,k} = \frac{1}{k-1}(k \cdot p^2 + \cos^2\theta \cdot A + \sin^2\theta \cdot B +$$

$$2 \cdot \sin\theta \cdot \cos\theta \cdot C - 2p(\cos\theta \cdot D + \sin\theta \cdot E) \qquad (16)$$

Next, the expression will be reformulated by isolating p:

$$DIST_{n,k} = \frac{k}{k-1}\left(\left(p - \frac{\cos\theta \cdot D + \sin\theta \cdot E}{k}\right)^2 + \frac{1}{k}\left(\cos^2\theta \cdot \left(A - \frac{D^2}{k}\right) + \right.\right.$$

$$\left.\left. \sin^2\theta \cdot \left(B - \frac{E^2}{k}\right) + 2 \cdot \sin\theta \cdot \cos\theta \cdot \left(C - \frac{DE}{k}\right) \right) \right) \qquad (17)$$

The squared parenthesis is always ≥ 0, implying that to minimize the distortion, the optimal p, denoted $\hat{p}$, will be chosen as (when knowing the optimal θ, denoted $\hat{\theta}$):

$$\hat{p} = \frac{\cos\hat{\theta} \cdot D + \sin\hat{\theta} \cdot E}{k} \tag{18}$$

Equation (18) can be reformulated as:

$$\hat{p} = \frac{1}{k} \sum_{i=n}^{n+k-1} r_i \cdot \cos(\hat{\theta} - \varphi_i) \tag{19}$$

which shows that p in fact is calculated as the arithmetic mean value of the orthogonal projections of measurement points along a line through the origin, and whose angle component θ is equal to $\hat{\theta}$. It is also easy to see that p cannot be negative.

In the sequel, it is assumed that p is selected as in equation (18), turning equation (17) into the following expression:

$$DIST = \frac{1}{k-1} \left(\cos^2\theta \cdot (A - \frac{D^2}{k}) + \sin^2\theta \cdot (B - \frac{E^2}{k}) + 2\sin\theta\cos\theta \cdot (C - \frac{DE}{k}) \right) \tag{20}$$

Next introduce H, I and J, defined as:

$$H = A - \frac{D^2}{k} \qquad I = B - \frac{E^2}{k} \qquad J = C - \frac{DE}{k} \tag{21}$$

leading to the following equation describing the distortion:

$$DIST = \frac{1}{k-1} \left(\cos^2\theta \cdot H + \sin^2\theta \cdot I + 2\sin\theta\cos\theta \cdot J \right) \tag{22}$$

The minimum can be found by calculating the derivative with respect to θ in equation (22), which will be done on the following analogous expression:

$$DIST = \frac{1}{k-1} \left(\frac{1 + \cos(2\theta)}{2} \cdot H + \frac{1 - \cos(2\theta)}{2} \cdot I + \sin(2\theta) \cdot J \right) \tag{23}$$

which, before calculating the derivative, can be simplified to:

$$DIST = \frac{1}{k-1} \left(\cos(2\theta) \cdot \frac{H-I}{2} + \frac{H+I}{2} + \sin(2\theta) \cdot J \right) \tag{24}$$

The derivative becomes:

$$\frac{dDIST}{d\theta} = \frac{2}{k-1} \left(\cos(2\theta) \cdot J - \sin 2\theta \cdot K \right) \tag{25}$$

where K is defined as:

$$K = (H - I) / 2 \tag{26}$$

Set $2\theta = q$ and the derivative $= 0$, yielding:

$$\cos(q) \cdot J = \sin(q) \cdot K \tag{27}$$

On the interval $[0, 2\pi[$, two possible solutions will occur for q:

$$q = \operatorname{atan}(J/K) + v \cdot \pi \qquad v \in Z \tag{28}$$

which in turn gives four solutions for θ:

$$\theta = 1/2 \cdot \operatorname{atan}(J/K) + v \cdot \pi/2 \qquad v \in Z \tag{29}$$

Define the function W through:

$$W = \begin{pmatrix} 0 & K > 0 \\ \pi & K < 0 \end{pmatrix} \tag{30}$$

(The case K=0 needs special treatment)

Now, the zeros of the derivative can be expressed uniquely in q in the following way:

$$q = \operatorname{atan}(J/K) + W \tag{31}$$

which is unique on the interval $[0, 2\pi[$.

This can be transformed to an expression in θ:

$$\theta = \frac{1}{2}(\operatorname{atan}(J/K) + W \pm \pi) \tag{32}$$

Each of these two values, substituted for θ in equation (22), will give the same value for DIST. This is due to the fact that it is possible only to determine the direction of a line through the origin of coordinates.

After DIST has been calculated with any of the solutions for θ, it is compared to a predetermined value, denoted THRESHOLD. If it is less than THRESHOLD, two candidate values for p are calculated by substituting each θ for $\hat{\theta}$ in equation (18). One of the p candidate values is positive, while the other is negative. The text following equation (19) states that p cannot be negative. Thus p is updated with the positive candidate value. The coordinates of the line will be updated with $(\hat{p}, \hat{\theta})$ just calculated (Broken line in figure 8).

If, on the other hand, DIST exceeds THRESHOLD, a new line will be created, starting from this last point that could not be added to the current line (The two solid lines in figure 8).

8.2 The Recursive Updating Scheme

Below follows a summary of the recursive calculation of the five partial sums (A,B,...,E).

$$\begin{cases} A_{n,k} = 0 & k = 0 \\ A_{n,k} = A_{n,k-1} + (r_n \cdot c_n)^2 & k = 1, 2, \ldots \end{cases} \tag{33}$$

$$\begin{cases} B_{n,k} = 0 & k = 0 \\ B_{n,k} = B_{n,k-1} + (r_n \cdot s_n)^2 & k = 1, 2, \ldots \end{cases} \tag{34}$$

$$\begin{cases} C_{n,k} = 0 & k = 0 \\ C_{n,k} = C_{n,k-1} + r_n^2 \cdot c_n \cdot s_n & k = 1, 2, \ldots \end{cases} \tag{35}$$

$$\begin{cases} D_{n,k} = 0 & k = 0 \\ D_{n,k} = D_{n,k-1} + r_n \cdot c_n & k = 1, 2, \ldots \end{cases} \tag{36}$$

$$\begin{cases} E_{n,k} = 0 & k = 0 \\ E_{n,k} = E_{n,k-1} + r_n \cdot s_n & k = 1, 2, \ldots \end{cases} \tag{37}$$

From Eqs. (33) - (37), J and K can be calculated as:

$$J_{n,k} = C_{n,k} - \frac{D_{n,k} E_{n,k}}{k} \tag{38}$$

$$K_{n,k} = \frac{1}{2} \left(A_{n,k} - \frac{D_{n,k}^2}{k} - B_{n,k} + \frac{E_{n,k}^2}{k} \right) \tag{39}$$

It is possible to express the calculation of J and K through recursive equations as well, but the prize to pay is high, since the recursive expressions become very complex.

Acknowledgments

We would like to thank Åke Wernersson and the staff at IKP, Linköping University for providing sensor data from their robot. Thanks also to Per Klöör, Andris Lauberts and Gunilla Borgefors, National Defence Research Agency for comments and suggestions.

References

1. Andersen C.S. *et al.* "Navigation using range images on a mobile robot", Robotics and Autonomous Systems, vol.10, pp.147-160, 1992.
2. Chatila R., Laumond J-P. "Position referencing and consistent world modelling for mobile robots", IEEE J. of Robotics & Automation, pp.138-145, 1985.

3. Cox, I.J. "Blanche: Position Estimation for an Autonomous Robot Vehicle", In: Autonomous Robot Vehicles, I.J. Cox, G.T. Wilfong (Eds.), pp.221-229, Springer-Verlag, New York, 1990.
4. Duda R.O., Hart P.E. "Pattern Classification and Scene analysis", John Wiley & Sons, New York, 1973.
5. Ekman, A. *et al.* "Incremental map-making based on range images", FOA Report C 30672-3.4 ISSN 0347-3708, Linköping, 1992.
6. Elfes, A. "Occupancy Grids: A Stochastic Spatial Representation for Active Robot Perception", pp.60-70, Proc. of the sixth Conf. on Uncertainty in AI, 1990.

Image Segmentation Improvement
with a 3-D Microwave Radar[1]

A. Siebert, J. Ostertag, B. Radig
Institut für Informatik, TU München, Germany
M. Rožmann, J. Detlefsen
Lehrstuhl für Hochfrequenztechnik (HFS), TU München, Germany
J. Bernasch
BMW AG, München, Germany

Abstract

Segmentation of intensity images is well known to be a complex problem. A single intensity image rarely provides all the information necessary for a correct segmentation. In this paper we describe how to improve segmentation results by fusing the intensity data with range data sensed by a microwave radar. The radar either gives a clue of where to partially repeat the segmentation process with different parameters or where to replace the intensity image segmentation data with the radar segmentation data. Practical results are obtained by applying the work to the vision system of an autonomous mobile robot where the goal is to identify and localize unknown objects.

1. Introduction

Extensive work on the segmentation of intensity images has proven for segmentation to be a hard problem [1]. Noise, occlusion, varying lightning conditions, and unknown optical properties of the world objects invariably lead to under- or oversegmentations that cannot be resolved using only a single intensity sensor. A correct segmentation, however, is an essential prerequisite for obtaining a correct image interpretation later on. An obvious way for improving the segmentation results is to implement different kinds of physical sensors. Especially useful are sensors who deliver range information. We therefore used a microwave radar to measure distance values that a video camera cannot provide. In contrast, other researchers have been working on different sensors like laser cameras and sonar systems that can also provide range data [2].

A suitable system architecture has to be devised to implement the sensor data processing in an efficient way. Our system is based on the blackboard paradigm where each sensor is represented by an independent knowledge source.

[1]This work was supported by the Deutsche Forschungsgemeinschaft as part of the joint project "Information Processing in Autonomous Mobile Robots (SFB 331)". It is a result of an interdisciplinary cooperation between the Lehrstuhl für Praktische Informatik and the Lehrstuhl für Hochfrequenztechnik.

The segmentation process is a significant part of a vision system which has been developed within the interdisciplinary research project at the Technical University of Munich that deals with information processing in autonomous mobile robots. The vision system's task is to localize and identify known and unknown objects which are obstacles to navigating a robot in an industrial environment.

2. System Components

A standard video camera with a resolution of 512×512 pixels, attached to a framegrabber, is used to acquire intensity images. The position and orientation of the camera are fixed and known. The lighting is a mixture of daylight and artificial light as it is typical for offices and industrial plants.

A 94 GHz 3-D imaging microwave radar has been developed at the Technical University of Munich [3,7,8]. Given the direction of the narrow radar beam, it measures the density of point scatterers within an observed volume cell. Such a volume cell can be considered to be a cube with a sidelength of 15 cm. The grid described by these volume cells can be displaced incrementally by 30 mm steps. The direction of the radar beam can be adjusted in azimuth an elevation within a broad field of vision with an angular accuracy of 0.125^0 which corresponds to 1.75 cm at a distance of 8 m. Note that the diameter of the radar beam is constant and that a scanning process is required to obtain a complete sensor image. In our experiments the scanning was done in stationary scenes. It is possible, however, to do the scanning in nonstationary scenes, e.g. if the radar is put on a slow moving robot. In this case only small variations of the elevation are allowed which results in a reduced field of vision.

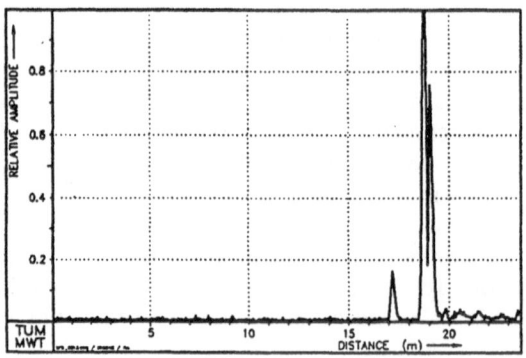

Fig. 1: Typical radar scan

Some preprocessing is required to deduce the range data, i.e. the distance, from the density of the point scatterers, thereby accounting for such effects like interference, multiple reflections, and absorption. Figure 1 shows a typical radar scan where the relative amplitude of the reflected beam is measured as a function of the distance. Here, the high resolution capability of the radar device becomes obvious. The first significant peak is caused by the object of interest, in this case a small cylinder, the second one, i.e. the double peak, is caused by the wall and a window behind the object of interest. The tiny peaks with a relative amplitude close to zero represent noise. Therefore, any signal below a certain threshold is suppressed.

Satisfactory measurements can be expected from objects whose surfaces are perpendicular to the radar beam. Otherwise, only a small but usually sufficient fraction of the scattered energy can be sensed. Here, the quality of the measurements depends on the reflection characteristics of the sensed object. Analogous to optical phenomena, Lambertian surfaces look equally bright in all directions while specular surfaces scatter the incident energy in one direction only. Our results show that in our application environment only an insignificant part of the range data is poor due to specular reflection.

The position and orientation of the radar are known. The position of the radar relative to the video camera is constant and known. Both sensors are installed next to each other on top of the mobile vehicle developed within our joint research project [2].

The image interpretation system [4] has been designed as a blackboard system [5]. This allows for each sensor to be modelled as a knowledge source. The knowledge sources observe the state of the blackboard and know in which situations they can contribute to the problem solving process. The blackboard stores all intermediate results. Besides the knowledge sources for each sensor there is a model knowledge source that contains models about the world objects and a control knowledge source that is in charge of the flow of control. In particular, the control knowledge source makes sure that the intensity image is the first sensor data to be evaluated.

3. Segmentation Improvement

A well known problem of multisensor systems is the large amount of data produced by the various sensors. In most applications using all data turns out to be prohibitive in terms of processing time. A data reduction strategy that cuts processing time has to be adopted.

An overview of the approach we take is shown in figure 2. It gives priority to the intensity data processing since video images can be sampled at high speed (up to 30 frames/second). A complete sampling and segmentation of the intensity image is done in any case. The microwave radar can be added in different ways [6], depending on the quality of the intensity image segmentation. A satisfactory segmentation is one that allows a successful object matching and recognition which in turn depends on the subsequent matching and recognition modul. In our blackboard system, this means that the video camera knowledge source determines which of the following actions the microwave radar knowledge source takes.

- The segmentation of the intensity image is good enough. This is the case if the objects under consideration can be recognized using only the intensity data. Then the microwave radar is not used at all.

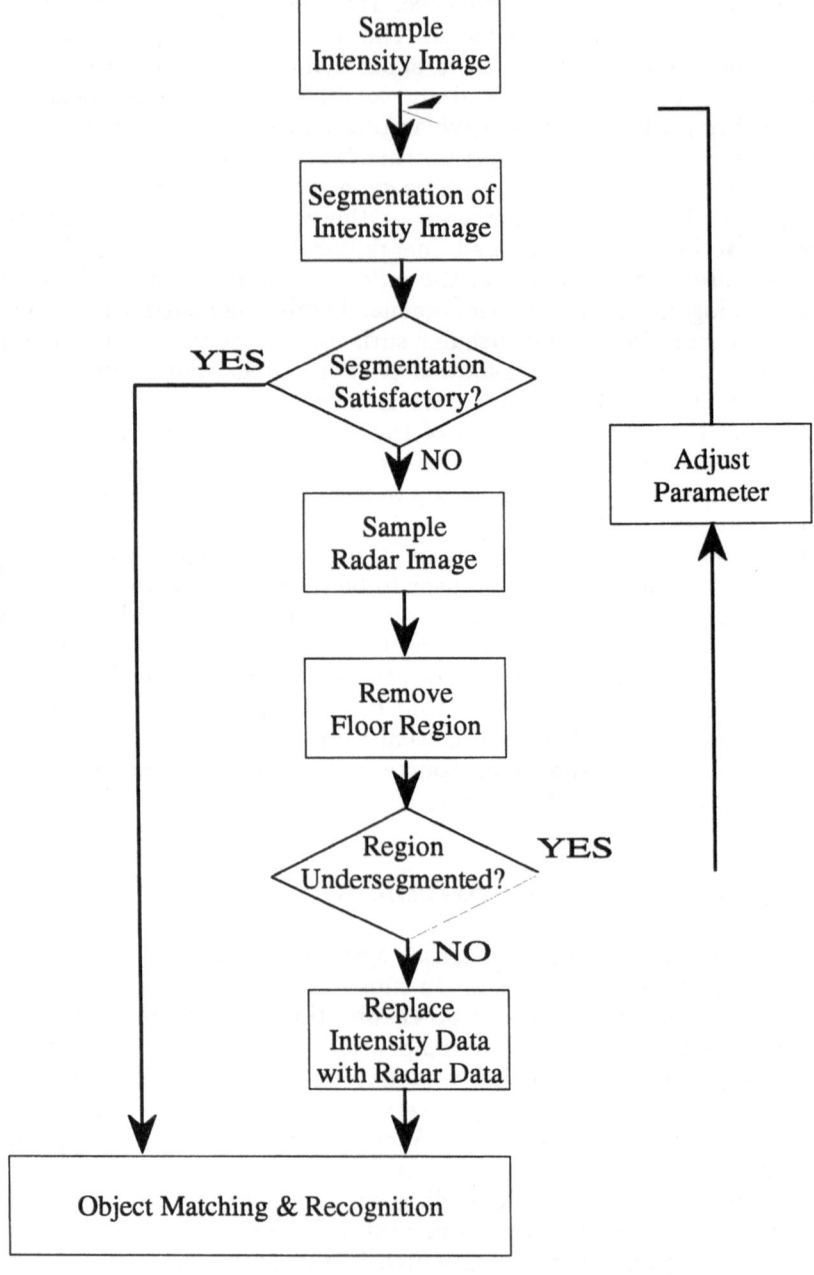

Fig. 2: Integration of intensity data and radar range data

- The segmentation of the intensity image is poor. Then the microwave radar is used according to the following strategy.
    1. A radar image is sampled. Regions, e.g. the floor, that can immediately be identified as not belonging to any object of interest are discarded.
    2. The radar identifies regions that are undersegmented. A region in the intensity image is undersegmented if the radar can detect step edges, i.e. depth discontinuities, within that region. Such regions are resegmented using automatically adapted parameters. In our scheme, this means that the intensity values of the region of interest are stretched over the maximum intensity range (256 grey values). This approach takes advantage of the high resolution of the intensity image. A necessary condition for it to work is that there are subregions which are significant different in terms of their grey values.
    3. If resegmentation does not lead to improved results, the radar data segmentation replaces the intensity image segmentation in the region of interest.

As can be seen, we use the microwave radar for a low-level and intermediate-level processing of pixels, edges, and regions. It should be clear that it is not rewarding in terms of processing speed and accuracy to build a high-level, symbolic description using only the radar data. Instead, the radar range data are meant to supplement and support the intensity data obtained from the video camera.

For the outlined strategy to work, the radar data and the video data have to be aligned pixelwise. Note that the alignment depends critically not only on the sensor geometry (which is known) but also on the distance of the object under consideration (which is measured by the microwave radar).

4. Results

Figures 3 and 4 show some typical results. A tool carriage that might be found in an industrial environment is to be segmented. Figures 3(a) and 4(a) show the unprocessed intensity images. The bad lighting condition is obvious. Figures 3(b) and 4(b) show the unimproved segmentations. As can be seen, the segmentations are poor since regions that belong to the floor or to the background wall are merged with parts of the carriage due to shadows. Recognizing the tool carriage is not possible. Figure 3(c) shows the improved segmentation obtained by resegmenting the largest region of figure 3(b). In that region the boundary lines of the tool carriage represent depth discontinuities that are identified by the radar. Figure 4(c) shows an improved segmentation obtained by replacing intensity data with radar data. Note that the boundary lines are fairly rugged. This effect occurs at step edges where only a part of the radar beam is reflected by the foreground object while another part is reflected by the background. Improved preprocessing techniques might be able to solve this problem. At this point, the resolution of the intensity image is obviously superior to the radar image's resolution.

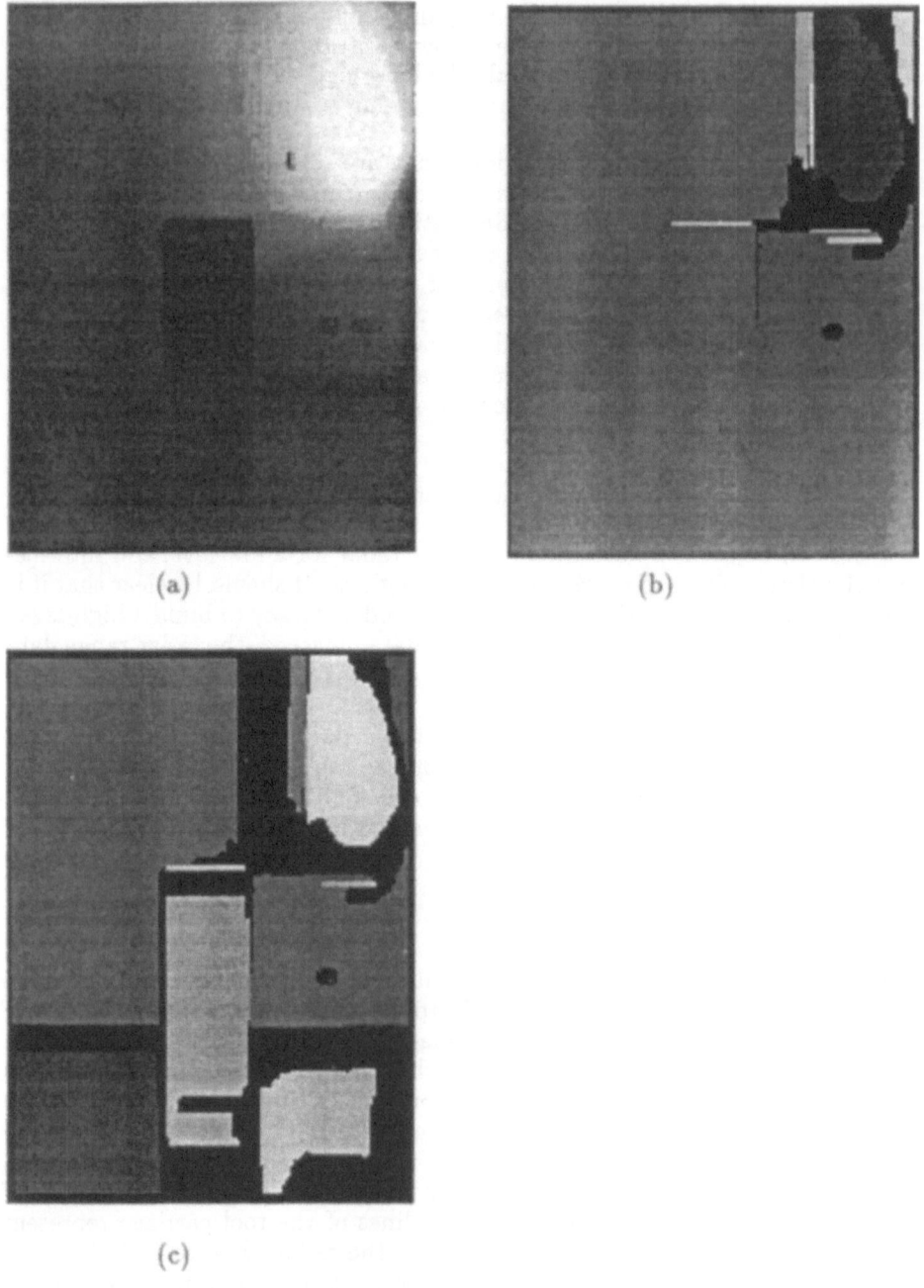

(a)

(b)

(c)

Fig. 3: Segmentation improvement by resegmentation (a) intensity image: the object of interest, in this case a tool carriage, is located in the center (b) unimproved segmentation (c) improved segmentation

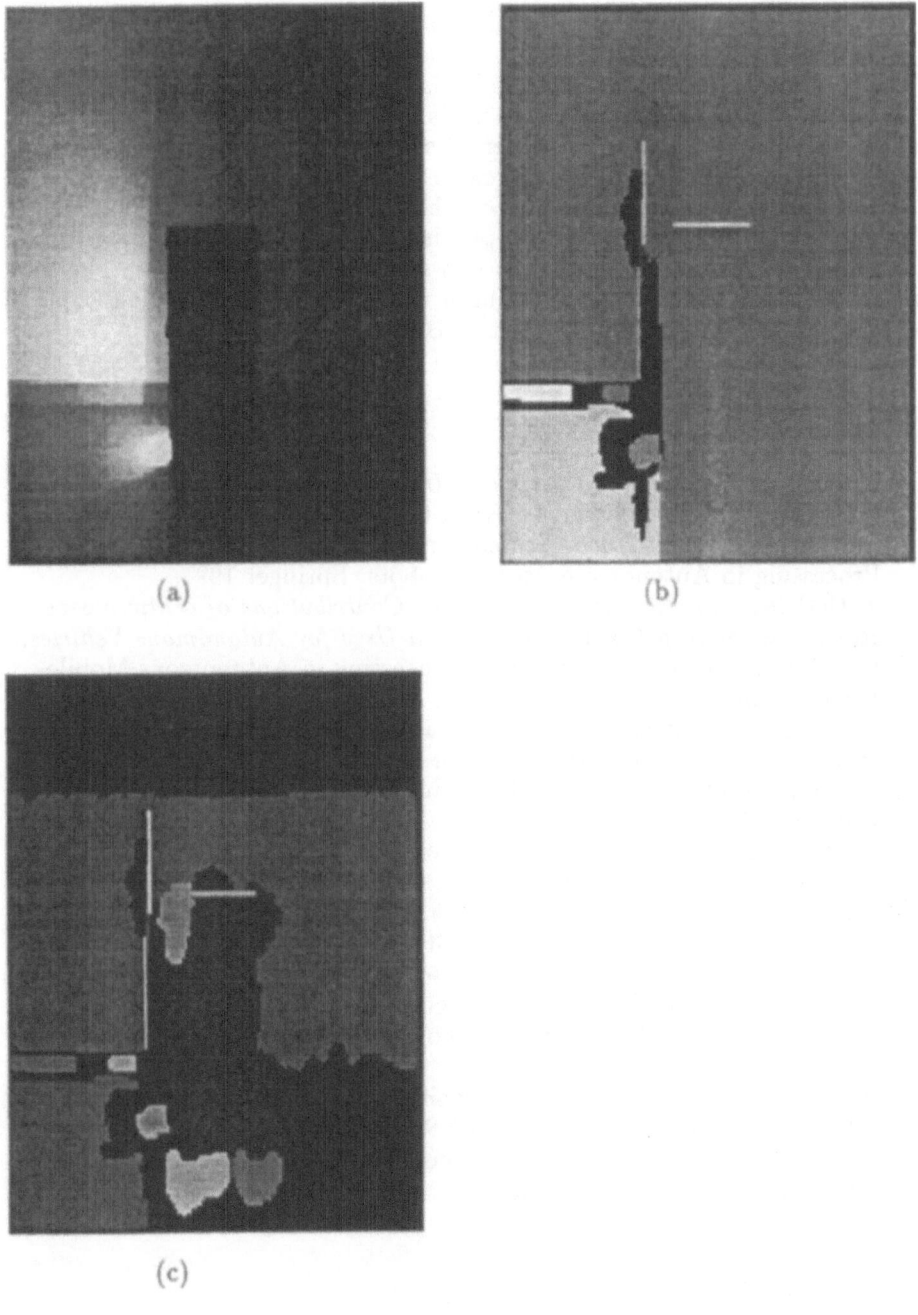

Fig. 4: Segmentation improvement by data replacement (a) intensity image: the object of interest, in this case a tool carriage, is located in the center (b) unimproved segmentation (c) improved segmentation

5. Conclusions

Clearly, the use of additional sensors leads to improved segmentation results. The microwave radar delivers range measurements that a video camera cannot provide. By integrating the different sensors in an efficient way, a correct segmentation was obtained in most cases.

Currently, more work is done to achieve a higher microwave radar performance. This work focuses on improving the preprocessing capabilities of the microwave radar and using it in non-stationary environments. The integration of other sensors that make up a multisensor system [2] is a topic of ongoing research in the joint project SFB 331.

References

1. H. Niemann, *Pattern Analysis and Understanding*. Springer 1990.
2. C. Fröhlich, F. Freyberger, G. Karl, G. Schmidt. *Multisensor System for an Autonomous Robot Vehicle*. In: G. Schmidt (ed.) Information Processing in Autonomous Mobile Robots. Springer 1991.
3. J. Detlefsen, M. Rožmann, M. Lange. *Contributions of a Microwave Radar Sensor to a Multisensor System Used for Autonomous Vehicles.* In: G. Schmidt (ed.) Information Processing in Autonomous Mobile Robots. Springer 1991.
4. P. Levi, O. Munkelt, B. Radig, R. Sattler. *An Application of Image Processing and Image Interpretation in Manufacturing Environments.* In: G. Schmidt (ed.) Information Processing in Autonomous Mobile Robots. Springer 1991.
5. R. Engelmore, T. Morgan. *Blackboard Systems.* Addison-Wesley, 1988.
6. J. Ostertag. *Verbesserung der Erkennungsleistung eines Bildverarbeitungssystems durch Auswertung der Daten eines 3D-Mikrowellenradars.* Diplomarbeit, TU München, 1993.
7. M. Rožmann, J. Detlefsen. *Millimeterwellen-Radarsensor für autonome, mobile Fahrzeuge,* in 6. GMA/ITG Fachtagung, VDI Berichte 939: Sensoren: Technologie und Anwendung, S. 101-106, 1992.
8. M. Rožmann, J. Detlefsen. *Environmental Exploration based on a Three-Dimensional Imaging Radar Sensor,* in Proceedings of the 1992 IEEE/RSJ International Joint Conference on Intelligent Robots and Systems, pp. 422-429, 1992.

A Vectorial Definition of Conceptual Distance for Prototype Acquisition and Refinement

Carlo Moneta, Gianni Vernazza, and Rodolfo Zunino
Department of Biophysical and Electronic Engineering
University of Genoa, Via Opera Pia 11A, 16145 Genoa, Italy

Abstract

This paper addresses the problem of matching symbolic descriptions of structured objects. The adopted methodology is the basic component of the decision-making control of a learning system for the acquisition and the refinement of prototypes of visual objects. A vectorial matching evaluation is proposed, as opposed to traditional scalar similarity-measures. This allows the final result of the matching process to account for both the structural similarities of the compared objects, and the information about the reliability of available descriptions. The decision-making mechanism based on such vectorial representation is also described. With regards to the system's overall flexibility, the advantage of connecting the matching output with the decision-making process via a common vectorial representation is highlighted.

1. Introduction

Symbolic matching in prototype refinement for a vision system is the main topic this paper deals with. Indeed, every research addressing symbolic model processing has the matching process as a central task, because it is on the basis of matching results that comparisons are evaluated, and decisions are made.

For this reason, a lot of work has been done to achieve a formal statement of *conceptual distance*, that is a methodology to match structured symbolic descriptions. For instance, [1] and [2] present a specialized analysis of the matching process to perform conceptual clustering; from the point of view of artificial intelligence and machine learning [3, 5], the problem of conceptual distance as a general basis to perform inductive learning tasks is addressed in [6].

When dealing with symbolic matching, one of the main concerns is the relevance of the context within which the matching process is being carried out.

The current lines of research show that *strict matching* (i.e. a matching approach which does not allow partial or flexible matching) is no longer an efficient technique to handle concept representations; several methodologies have been developed to allow *flexible matching* [7], so that concepts' descriptions may also drive context-dependent matching tasks. From another point of view, traditional matching techniques result in a scalar quantity (say, a distance), but the need may arise for some more information-rich outcomes of the matching process. In this sense, uncertainty may be present in some of the compared objects' features, or fuzziness may characterize their descriptions; a well-suited matching technique should take into account all these possibilities, and the results of the process should also be able to reflect the varying situations.

This paper presents a syntactic approach to flexible measurement of conceptual distance; the adopted technique is defined as *flexible*, first because the structure of concepts' descriptions may change while the system learns, and then because every involved concept is also provided with a *reliability measure* characterizing its own description. The latter is an important and useful feature, when dealing with dynamic concept acquisition, because it allows one not only to improve the knowledge embedded in the system, but also to assess the degree of confidence that can be assigned to such knowledge.

In the HURBINEK system, such techniques are implemented for the acquisition and refinement of symbolic models (prototypes) for object recognition; such system operates within a larger multisensorial environment for autonomous land-vehicle control [8, 9]. In Section 2, the structure of the prototype refinement system (HURBINEK) is briefly outlined, the basic conventions for knowledge representation are defined, and the overall operation is sketched.

In particular, the system's features that mostly influence the implementation of the matching process are pointed out. Section 3 addresses the specific problem of symbolic matching; basic issues are discussed, such as the influence of data representation on the evaluation of matching, and a special attention is focused on how uncertainty is handled and accounted for by the matching process. Moreover, the vectorial representation of the matching results in a 3-D space is described, showing the usefulness of such representation in relation to the consequent decision-making process. Some concluding remarks are made in Section 4.

2.   Structure of the Hurbinek System

The HURBINEK[1] system is a learning system dedicated to the acquisition and refinement of concepts' prototypes.

In general, prototypicality represents a major issue in recognition systems for sensorial data understanding, especially in hierarchically-structured knowledge-based systems for scene interpretation. For example, suppose that a

---

[1] is not an acronym; it is the name of a character drawn from Italian literature, representing a child who is not able to speak, but wants to learn.

top-down expectation-driven recognition process predicts that an object (say, a house) is to be found in the scene being examined. If no additional information about such object is supplied, its prototypical representation should be obviously used first.

In this sense, a prototypical representation must be general enough to cover a wide variety of instances, and, on the other hand, must be specialized enough to speed up the recognition process by driving the interpretation of sensorial data properly. For this reason, a methodology for automatic extraction and refinement of prototypical descriptions seems to be quite useful.

The HURBINEK system operates at a symbolic level of data representation; a low-level processing phase is assumed to be performed on sensorial data, so that raw sensor signals can be compiled into a symbolic form made of descriptive features. This can be accomplished by a data-fusion integrated system like the one described in [9].

2.1 Data Representation

The basic element of HURBINEK's knowledge representation is the *node* or *item*; there exists a node in the system for each defined object and concept.
Nodes are organized hierarchically, and the resulting structure resembles quite closely a semantic network [10, 11]. The general structure of a node is the following:

- name;
- upper link (hierarchical relation to the parent node);
- lower links (hierarchical relations to sons);
- relational descriptions (*features*), in the form of attribute-value pairs;
- structural descriptions (*predicates*), describing relations among the node's fea tures, and allow the representation of structured objects.

Excluding the name and the hierarchical links, a relialibility measure is associated with every feature (both attribute and value may have a different reliability measure) and predicate. The following could be an example of a node:

```
( CHURCH
        is_a BUILDING
        sons (CATHEDRAL PARISH CHAPEL)
                [ size (100) ..... large (80) ]
                [ entrance (100) ..... large (100) ]
                [ bell-tower (90) ..... tall (100) ]
                [ windows (100) ..... small (90) ]
                . . . )
```

Predicates will be omitted in the following for the sake of simplicity, but they can be treated in ways similar to those described hereafter. The quantities associated with node's attributes represent the percentage of sons that share those attributes; the quantity attached to each value indicate the sons' percentage sharing that attribute-value pair.

In general, these features can be regarded as links connecting a node to other nodes in the network; in this sense, the quantities associated with links represent a sort of *statistical weights* describing the links' reliability. In the following, such quantities will be denoted as *statistical quantifiers* (SQ_a for attributes and SQ_v for values). The following relation is defined as the overall weight of a link :

$$SQ_f (link) := SQ_a(link) * SQ_v (link)$$

2.2 Data Processing

The input datum (INPUT) supplied to the system is represented by a collection of features that is homogenous with the network nodes' structure. Inputs are the symbolic results of lower-level processing of sensorial data, and are supplied to the system one at a time. In this sense, the learning process can be classified as *incremental.*

The first task the system must perform is the classification of each incoming input. This process results in the identification of the network's nodes that are most similar to INPUT, for their own properties or description. This set of nodes is called the INPUT's 'family'; the lower those nodes are in the network hierarchy, the more specialized is the output of classification.

In the training phase, the teacher's role is to confirm the classification proposed by the system. When confirmation has been achieved, INPUT is inserted in the network, and the actual learning process, .i.e. prototype refinement, is activated. This is a summary of the basic system's operation:

1. Get the incoming 'INPUT' description (from low level processing)
2. Classify INPUT
3. Achieve confirmation/refusal (from the teacher)
4. If refused, repeat pass 2. trying other classif. (backtracking)
5. (If confirmed) insert INPUT in the network
6. Activate generalization processes on INPUT's 'family'.

The classification process is performed in a top-down fashion, starting from high nodes in the hierarchy, and getting down to lower ones. At each level, INPUT is compared to the network nodes to decide which one among them is the most promising for more specialized classifications. In particular, the decision must be made as to:

• *stop* classification at the current level and propose it to the teacher;
• return *fail* due to mismatch, and force backtracking at higher levels;
• *continue* the classification process down to lower (most promising) nodes.

The assessment about *promising nodes* is obtained from the results of the matching process. Those results must accomplish the following requirements:

• return a similarity measure between INPUT and the network nodes;
• address the decision to be made about classification in accordance with the current *family* situation.

In this sense, this decision-making mechanism requires a rather sophisticated matching process; in general, there is not a one-to-one correspondence among the descriptive structures (attribute-value pairs) of the compared objects; moreover, the reliability measures associated with descriptive links must be taken into account. The following Section describes the syntactic approach adopted to cope with these problems in the matching evaluation.

3. Matching Process

This Section will be divided in two subparts: first, the definitions and the procedures for actual matching computation are introduced; then the methodology adopted to make decisions on the basis of such matching results is shown.

3.1 Matching Computation

The matching process involves two objects whose descriptive structures have been presented in Section 2: the INPUT node, obtained from symbolic compilation of low-level data processing, and a network node, which will be denoted as the *GOAL* node. Figure 1 presents a useful representation of the matching environment. In this schema, an element of a set represents a descriptive link (feature). Therefore, the following subsets can be defined:

- IMA (Input Mismatch Area): the set of INPUT's features not shared with GOAL;
- GMA (Goal Mismatch Area): the set of GOAL's features not shared with INPUT;
- MA (Matching Area): the set of features shared by INPUT and GOAL.

A scalar quantity is associated with each of such subsets, representing the area's relative significance with respect to the others:

- IMM (Input Mismatch Measure) - associated with IMA;
- GMM (Goal Mismatch Measure) - associated with GMA;
- M (Matching-figure) - associated with MA.

Such quantities must take into account both the *qualitative* and the *quantitative* aspects of the related information; in other words, both the number of features present in a set and their associated reliability measures must be considered in the whole evaluation process.

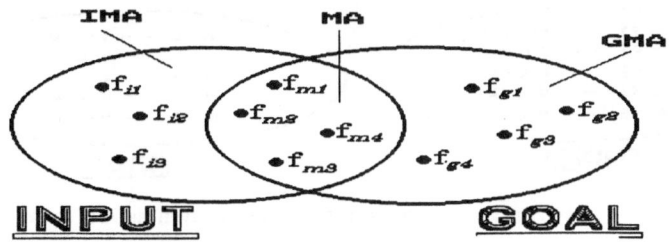

Fig. 1: Significant areas in matching evaluation

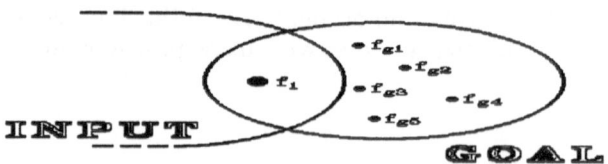

Fig. 2: Different weigths influence matching evaluation

For instance, consider the example represented in figure 2:

All the f_g's have very low reliability measures, whereas f_1 has a very high reliability measure. In this case, evaluating GMM greater than M, due to the different number of features in the two sets, could be incorrect because the contribution of uncertainty should also be taken into account.

On the other hand, consider the situation depicted in figure 3.

In this case, the two compared objects share a common structure (set of attributes) as opposed to their different actual descriptions (set of values); if reasoning by analogy is being performed by the system, this situation could be considered as a *good matching* one, disregarding the nodes' specific characteristics (values).

The general problem of arranging the similarities of structural descriptions together with the relevance of reliability measures has been approached by the definition of the following two quantities:

- VOLUME: accounts for the number of features constituting a set;
- WEIGHT: reports on the the relative significance of the involved features.

The related expressions are given below:

$$(VOLUME) \rightarrow VOL(x) = \frac{\# feats(x)}{\# feats(node)}$$

$$(WEIGHT) \rightarrow WGT(x) = \frac{1}{\# feats(x)} * \frac{\sum_{feat(x)} SQf(x)}{\sum_{feats(node)} SQf(node)}$$

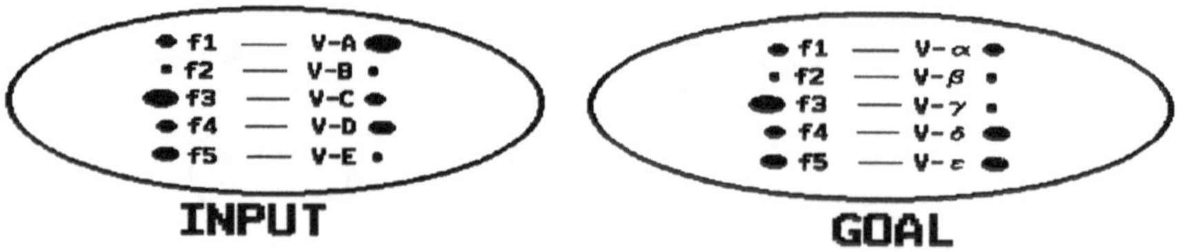

• f1 — V-A	• f1 — V-α
• f2 — V-B	• f2 — V-β
• f3 — V-C	• f3 — V-γ
• f4 — V-D	• f4 — V-δ
• f5 — V-E	• f5 — V-ε
INPUT	**GOAL**

Dot size is a visual representation of reliability measure

Fig. 3: Different structure influence matching evaluation

Volume and Weight are combined in a linear relation to work out the required matching figure:

$$IMM = K_1 * VOL(IMM) + K_2 * WGT(IMM)$$

$$GMM = K_1 * VOL(GMM) + K_2 * WGT(GMM)$$

$$M = K_1 * VOL(MA) + K_2 * WGT(MA)$$

Parameters are arranged so that:

$$0 <= x <= 1 \qquad x << \{ M, IMM, GMM \}$$

In the current implementation, K_1 and K_2 have been set to the same value (0.5), but the above definitions also allow one to modify the overall matching criterion by simply adjusting those parameters.

For example, if the structural similarity of two nodes is to be privileged over their descriptive appearance, K_1 will be increased and K_2 descreased; the inverse operation will be done if preference is assigned to certainty factors in the measurement of objects' similarity. In general, the application domain will determine the proper choice for the set of values $\{ K_1 , K_2 \}$.

3.2 Matching-driven Decision-making Process

The results of the computation of the above defined matching quantities are arranged in a 3-D vector:

$$m = (M, IMM, GMM)$$

As shown in figure 4, this vector addresses points in the $[0,1]^3$ cube. This involves the definition of several regions in this cube, each of them corresponding to a decision to be made by the system.

In the following, some examples are shown of the association of cube-regions with decision-making outcomes:

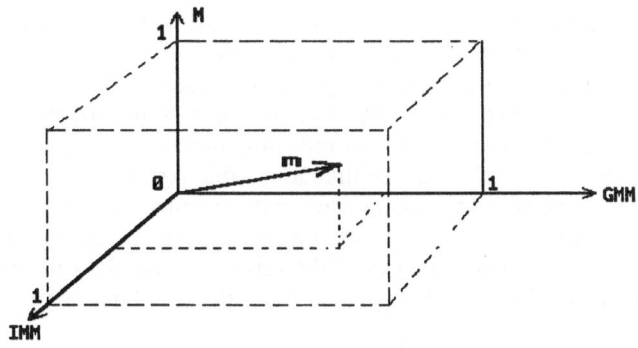

Fig. 4: The cube addressed by the 3D matching vector

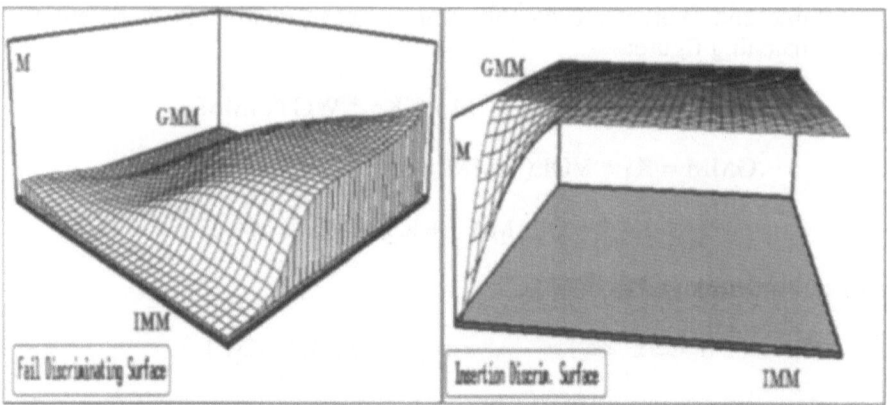

Fig. 5: Discriminat surface for decision-making process

- great M, small IMM, small GMM: this corresponds to a region close to the M-axis, and represents a remarkable similarity between INPUT and GOAL (see figure 5); in this case, the proposed decision is to *stop* the classification at this level, because a *brother* has been found;
- great M, great IMM, small GMM: INPUT has a more detailed description than GOAL; however, the great M value suggests that INPUT could be probably inserted in the network under some of GOAL's descendants: the system's decision will be to *continue* classification with some of GOAL's children;
- small M, great IMM, great GMM: this configuration addresses a situation of possible mismatch between INPUT and GOAL, and in general suggests to return *fail*; backtracking will be forced at some hierarchy level higher than GOAL's one.
- good M, small IMM, great GMM: the small value for IMM shows that INPUT is poor of information with regards to GOAL's description, so that continuing classification seems not very promising; however, the good value of M indicates that the two objects do have a significant similarity: the resulting decision will be to *stop* classification at the current level and insert INPUT as a *brother* of GOAL (insert-aside).

The vector-driven decision-making process can be easily implemented if the cube regions associated with system's decisions are bounded by suitable *discriminating surfaces*. In the present implementation, such surfaces have been defined in an analytical form, so that adjustable parameters can be introduced to control the surfaces' shape and consequent system's decisions.

Some examples of discriminating surfaces are presented in figure 5 and 6. For instance, the analytical form of the fail surface is the following:

$$FAIL(x,y) = G_0(x,y) + F_0(x,y) * H_{GM}(x,y) * H_{IM}(x,y) * H_{FAR}(x,y)$$

where :
- $G_0(x,y) = A_0 * \exp[-(x^2 + y^2) / V_1]$; {Gaussian centered in (0,0)};
- $F_0(x,y) = 1 / \{1 + \exp(- [sqrt(x^2 + y^2)-r_0] / V_2) \}$; {rotated Sigmoid centered in (0,0)};

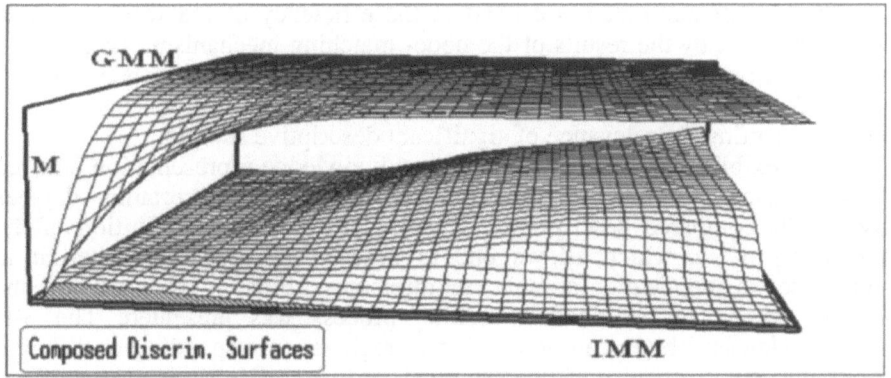

Fig. 6: Simple discriminating surfaces composition

- H_{GM}, H_{IM} are functions that model the slope of F_0 close to the GMM and IMM axis;
- H_{FAR} is a Gaussian function that models the F_0 slope far from the origin.

The major advantage of this approach lies in the integration of a vectorial representation of matching results with a flexible decision-making methodology. The latter draws its conclusions from matching outputs, but is independent from the method used for matching computation.

This means that the 3-D cube can model the 'procedural' methodology adopted to control the system's behaviour; moreover, the definition of analytical surfaces does not require the use of complex sets of meta-rules to drive the decision-making process. On the other hand, changes can be made to the basic matching evaluation, with no need for modifying the matching results processing.

In conclusion, the vectorial representation of m seems to be a suitably flexible way to map the matching results directly into the decision-making module of the system.

4. Concluding Remarks

A syntactic approach to symbolic matching was presented, within a learning system (HURBINEK) for visual prototype acquisition and refinement. The system operates at a symbolic level of data representation, and processes incoming inputs incrementally.

The information about the descriptions of models (prototypes) is embedded in a hierarchical relational structure similar to a semantic network, with additional characterization about the degree of reliability associated with each descriptive component.

Data supplied by low-level processing are first classified in a top-down fashion to find out the nodes in the network which are closest to the input description; then, the knowledge refinement process is activated, which performs modifications of prototypes' descriptions by means of generalizations and clustering techniques.

The classification step is critical to the efficiency of the whole process, and is controlled by the results of the node- matching mechanism. Therefore, a vectorial definition was proposed to perform symbolic matching between two structured descriptions. The matching vector m has three components, each of them representing the relevance of significant descriptive subsets.

The flexibility involved by the adopted knowledge representation allowed a differentiated treatment of structural similarities and uncertainty degrees between the compared descriptions. This lead to a flexible definition of the components of m, which allows one to modify the matching criterion in compliance to application-domain requirements. Finally, the use of the matching vector in the decision-making process was presented. The 3-D subspace addressed by m can be divided in regions, corresponding to different actions that the classification process must perform. Discriminating surfaces bounding such regions have been defined in an analytical form, and parameters can be suitably tuned to control the actual classification behaviour.

The advantages of this approach have been discussed, evidencing how the vectorial approach allows one to have the matching process independent form the decision task, and at the same time to keep their integration safe and easy.

References

1. Hayes-Roth F, McDermott J, "An interference matching technique for inducing abstractions" - Comm. of the ACM May 1978, vol 21-5.
2. Ben-Bassat M, Zaidenberg L, "Contextual template matching: a distance measure for patterns with hierarchically dependent features" IEEE Trans on Pattern Analysis and Machine Intelligence PAMI - vol 6-2, March 1984.
3. Michalski R S, Carbonell J G, Mitchell T M, Eds., "Machine Learning: an Artificial Intelligence approach" - vol. 1 Palo Alto, CA: Tioga 1983.
4. Michalski R S, Carbonell J G, Mitchell T M, Eds., "Machine Learning: an Artificial Intelligence approach", vol. 2 Palo Alto: Morgan-Kaufmann 1986.
5. Mitchell T M, Michalski R S, Carbonell J G, Eds., "Machine Learning: a guide to current research" - Amsterdam, The Netherlands, Kluwer 1985.
6. Kodratoff Y, Tecuci G, "Learning based on conceptual distance"; IEEE Trans. on Pattern Analysis and Machine Intelligence, vol 10-6 Nov 1988.
7. Bergadano F, Matwin S, Michalski R S, Zhang J, "Representing flexible concepts in knowledge-base systems" - MLI 5-88, TR 12-88, G. Mason University, VA Oct 1988.
8. Capocaccia D, Damasio A, Dellepiane S, Regazzoni C, Vernazza G, "Dynamic evaluation of multiple sensor for 3D obstacle detection" - Conf. on Time-varying image processing and moving object recognition, Florence Italy - May 1989.
9. Merialdo P, Pecollo P C, Regazzoni C, Vernazza G, Zunino R, "Integration of territorial maps in the vision sytem of an autonomous land vehicle" - Int.Conf.Intell.Aut.Systems, Amsterdam, The Netherlands, Dec 1989.
10. Woods W A, "What's in a link: foundations for Semantic Networks" - in Bobrow D G, Collins A, Eds., "Representation and Understanding" - Academic Press NY 1975.

Distributed Knowledge-based Systems
for Integration of Image Processing Modules

C. Regazzoni
Department of Biophysical and Electronic Engineering
University of Genoa
Via all'Opera Pia 11A, 16145 Genova, Italy

Abstract

Knowledge Based Systems (KBSs) and their application to Image Processing problems is described. The focus is on Artificial Intelligence techniques for knowledge representation that have been used for Image processing purposes. Production Systems and Semantic Networks are addressed as the most widely used techniques. Then, distributed approaches to KBSs, which have been of growing interest in the last few years, are considered, together with uncertainty management techniques, which are a major issue when dealing with noisy data and incomplete a-priori knowledge. KBSs are also classified into two main categories depending on the type of knowledge representation: object centered and process centered KBSs. Existing Image Processing applications of KBSs, of both the centralized and the distributed types, are briefly reviewed.

1. Introduction

An information Processing (IP) system S consists of a set of rules that perform symbolic and numerical inference's processes necessary to solve a specific problem. At the most general level, a problem is regarded as the task of finding a solution O, starting from input information, I (see figure 1 up). Knowledge-Based Systems (KBSs) make explicit two additional parts contained inside a system S: a-priori knowledge K (i.e., the part of the rules that is common to each method able to find the solution O), and the inference method itself, M (see figure 1 down).

Information Processing systems where knowledge is not clearly separated from the method M are usually referred to as algorithms, A. One could say that knowledge is compiled inside algorithms, i.e., it is implicitly represented, while the M part of a KBS has the role of interpreting knowledge K by selecting those chunks of it that are useful to produce a solution O, starting from information I.

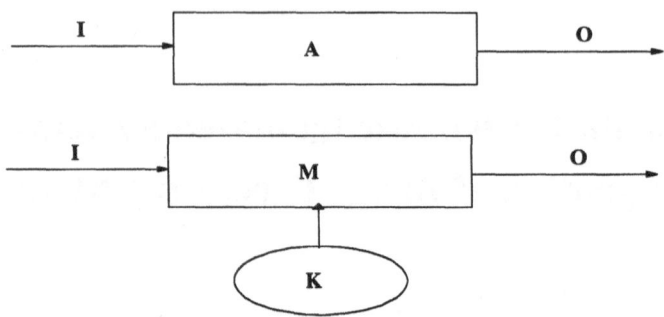

Fig. 1: Algorithmic Information Processing system (up); knowledge based information processing system (down)

From this description, it follows that a KBS can be characterized by a higher degree of adaptability. Moreover, if knowledge K has a modular structure, it allows modifications to the system to be easy performed: when a new application domain is examined, i.e. the class of problems on which the KBS operates is extended, it is easier to insert/change/delete the knowledge base without modifying the method M. In this sense, the separation between K and M allows both knowledge-engineers and users to better understand the functionality of a KBS in respect to what happens for a normal IP system. Moreover, as complex systems imply high development costs, it is vital to use their problem-independent parts for other applications.

However, generality often means poor computational efficiency. Efficiency problems are not an easy task to be solved. Compiled knowledge embedded inside algorithms surely allows, for example, a better problem-dependent allocation of memory resources and a faster processing time than KBSs. Especially for systems used in applications requiring high performances, it is often important to distinguish between different functionalities necessary to solve a problem. Therefore, a further requirement for a good Knowledge-Base K is the capability to evaluate the functionality of particular chunks during the solution of a specific problem, i.e., K must be self-selectable. Consequently, the method M must be able to select useful knowledge for solving a problem related to the processing of a specific class of input information I.

This approach leads to the use of Hybrid Systems (HSs) which include both algorithmic and knowledge-based parts. In HSs the work is performed at two levels, which correspond to two different types of knowledge representation in the Knowledge-Base: the tactical level and the strategical level. The tactical level is essentially algorithmic: specific processing steps which are performed in a fixed way are represented as not-interruptable operations. Specific solutions can be applied to efficiently code the particular knowledge necessary to such tasks. However, many of such tactical algorithms can be parametrized, i.e., their results may vary by appropriately controlling some input parameters. Moreover, the obtained result often changes depending on both the sequential order of activation of the algorithms and the different choice of alternative methods. The strategical level takes care of such problems, by using explicit knowledge attached to each algorithm in order to both set parameters and fix sequences of operations to be performed. The main difference between KBSs and HSs is that in HSs only a part of the knowledge used during the process execution is explicit, whereas compiled knowledge cannot be accessed, but only results of applications of such knowledge to data are available. In this case, the solution of a problem may consist of an enlarged output space, which also includes the selected sequence of algorithms, and the related set of chosen parameters, as a part of the actual process results.

The advantage of using KBSs and HSs becomes more evident when the size of the problem to be solved increases. A problem and the related system S may be large for two causes: the presence of a large set of inputs and/or a great difference between the input space and the solution space O. The most direct solution to such problems is to split the solving system into several subsystems. Even though modularity is often claimed to naturally arise from a hierarchical composition of different algorithms, it often seems that the resulting complexity cannot be managed without using tools allowing for an explicit representation of the a-priori knowledge related to the decomposition model and to the interdependencies between the input and the output of each subsystem. Moreover, it is well-known that the performance of a complex hierarchical system is limited by the worst subsystem in the chain. To overcome this drawback, the possibility of backtracking local errors on the basis of a deeper knowledge of the process is mandatory. HSs systems are able to maintain an high-level explicit representation of the different parts of a complex system by limiting efficiency loss. Moreover, they allow to easily individuate the causes of poor system performances, which can be often created with bottlenecks represented by specific algorithms.

Image Processing Systems (IMPs) use as input two-dimensional signals provided by different physical sensors, such as TV cameras, Infrared devices, Laser Range Finders, Sonars, Biomedical Tomographs, Synthetic Aperture Radar, etc.. IMPs are very complex systems for two main reasons: first, images are signals containing a lot of information coming from multiple external sources, and stored as several hundreds of Kbytes. Secondly, the problems they have to solve imply an output space relatively far from the numerical signal level. An image can be used to recognize and locate objects in a 3D scene for surveillance purposes or for autonomous guidance of vehicles, as well as to make a diagnosis about human health, or to locate crops on the Earth's surface. In all these cases, Image Processing tasks involve the solution of inverse problems, i.e., deducing the causes associated with numerical signals. To this end, it is natural to adopt a divide-and-conquer strategy, i.e., to progressively reduce the complexity of a problem through its decomposition into multiple, hierarchically organized processing steps. In addition to these considerations, one has to take into account that the current trend of research and industrial applications tends to see IMPs as part of larger systems based on multiple, physically distinguished sensors. Such systems are referred to as Data Fusion Systems (DFSs) [12, 32] and have the major advantage of offering a more complete basis for the solution of a problem, thanks to the different information they are able to process. DFSs are one of the main reasons for the necessity of investigating Hybrid Systems, as they are often proposed as the solution to very time and industrial performance-constrained problems. From these considerations, it follows that IMPs can take advantage of techniques and methodologies developed in the context of Artificial Intelligence for KBSs, and, more recently, of the consequent developments provided by Distributed Artificial Intelligence (DAI), which solves, in a systematic way, the issues related to the decomposition of a problem into a set of partially correlated sub problems.

The paper is organized into five parts: in section 2, classical KBSs and their applications to Image Processing are described. In section 3, new trends derived from DAI concepts are pointed out. section 4 provides a brief overview of existing applications of KBSs to Image Processing. Conclusions are presented in section 5.

2. Knowledge-Based Systems for Image Processing

The Knowledge required to solve Image Processing problems can be classified into two main categories: knowledge about the process itself and knowledge about objects. In both cases, knowledge can be either procedural or declarative.

Declarative knowledge deals with concepts, their attributes, and their relationships. Concepts can be either prototypical or instantiated. A prototypical concept is part of a-priori knowledge K, while an instantiated concept is contained in the solution space O. Procedural knowledge usually refers to the functionalities of a system, i.e. the sequence of actions that allows a system to progressively produce the solution O by combining prototypical and instantiated concepts. The difference between procedural and declarative knowledge has mainly a practical significance. The classification of such kinds of knowledge depends on the method M, on the selected representation of knowledge K, and on the solution O. It is often possible to generalize a problem by transforming its procedural knowledge into declarative knowledge, e.g., by representing a functionality as a concept and by applying procedural knowledge (at a higher abstraction level) to the obtained representation. HSs are an example of this approach: in HSs the explicit representation of declarative knowledge about employed algorithms allows the method M to apply procedural knowledge related to the strategic level. In many KBSs procedural knowledge concern with specific actions to be applied in a given application-domain, and, consequently, it is referred to the tactical level. This becomes clearer if one considers two kinds of systems generally used to implement KBSs: Production Systems and Semantic Networks. While describing such systems, another important point will be addressed, i.e., how uncertainty about the possible status of a system can be represented and propagated. Uncertainty management techniques will be summarized in the last part of this section.

2.1 Production Systems

In *Production Systems*, [1] knowledge K contains a set of rules. Each rule is a condition-action pair, whose left part considers the solution space O by evaluating if a given situation occurred. All rules satisfying the condition contained in the solution space are included in a conflict-set, and one of them is chosen according to a criterion (e.g., a matching measure). The right part of the selected rule is executed, so modifying the solution space O. These steps constitute the basic evaluation-and-execution cycle, which is applied until a termination condition is satisfied.

In the vocabulary of Production Systems, the inference method M is called the *interpreter*, and the solution space O is called the *database* (or, in a partially correct way, the *blackboard*). A deeper discussion is required by knowledge K.

Simpler Production Systems list rules without explicitly representing their relationships and related degrees of confidence. In such cases, a distinction between declarative and procedural knowledge is difficult. One may say that rules embed prototypical concepts regarding both objects and the process itself. The inference method is so general that the functionalities of a system could be derived from the status of the solution space during evaluation-and-execution cycles. This fact explains the strength and weakness of this kind of systems when applied to image processing.

On one hand, an apparently very general inference method is available which allows one to separately use different rules concerning with a specific problem. On the other hand, interferences among problem functionalities and objects during reasoning are difficult to manage if connections between consequent rules are not explicitly represented, and this may cause uncontrolled functioning.

According to the laws used to propagate uncertainty inside a system, Pearl classifies [2] Production Systems as Extensional Systems, as they usually attach a truth value to each rule and compute the uncertainty of the conclusion obtained by applying the rule itself as a function of the truth values attached to the conditions. In this way, one can associate an uncertainty value with each fact established by firing a rule. However, it becomes difficult, during the design phase, to keep trace of all possible paths, i.e., sequences of activated rules followed by a system.

Therefore, to design a robust system should imply the possibility to extensionally tracing all possible paths. This is an NP-complete problem which cannot be faced when the input dimensionality and the complexity of a task increase.

Consequently, due to lack of robustness, it may occur that the system reaches some status that cannot be managed by any rule, so causing the system to stop with no well-assessed conclusion. Nevertheless, it is difficult to debug the system knowledge completely, to assure that its performances will not degrade abruptly when input data are slightly varied.

This is an undesirable effect for all systems, especially for IMPs (see also the concept of ungraceful degradation, and regularization theory [10,17]), because one should be able to obtain some responses from the system (e.g., a partial recognition), possibly characterized by different degrees of confidence, depending on the kind of knowledge that has been used to reach such responses.

2.2 Semantic Networks

Semantic Networks [4] are another powerful method for representing knowledge. Both prototypical and instantiated concepts are represented as nodes of a graph. Links connecting nodes are associated with relationships between concepts. Depending on the semantic contents of concepts and links, such networks may be used to represent either procedural or declarative models of a problem. In Production Systems, the only way of maintaining consistency between multiple rules dealing with the same object or algorithm is to use an identical name for the object or algorithm in different rules. In this way, the same part of the solution may be accessed or modified by different rules. When using semantic networks, a higher degree of generality is possible: a single node may be used to represent a single concept in the knowledge-base, and multiple links can be used to specify possible relationships between the concept and other ones.

Semantic Networks are a basic tool for representing knowledge in a declarative format, and must be associated with a predicate calculus able to select interesting nodes (i.e., a method M) [5] and to perform inferences necessary to allow a KBS to produce a solution O.

This index scheme may be either centralized as a set of rules that form a Production System working at a higher representation level (i.e., on general concepts organized into a network and not on problem-specific concepts) or distributed, i.e., each node of the network is associated not only with declarative knowledge but also with chunks of procedural knowledge that can be applied locally.

If the inference method M used together with a Semantic Network is a PS, it has to deal with a smaller number of concepts than for classical PSs, thanks to the higher degree of generalization to which its rules operate. Consequently, a system made up of an SN and a PS is generally more robust than classical PSs, and allows one to design representations within a sufficiently wide scope. This makes Semantic Networks, as suggested by [6], suitable tools for developing *epistemological adequate* IMP systems, i.e., systems that express all facts actually used to search for causes associated with an image. However, this does not warrant that each SN plus PS is a good IMP because the reduced number of inference paths gained thanks to the highest generalization degree only limits, without eliminating, the possibility of an abrupt degradation of performances.

If a distributed index scheme is used, Semantic Networks are usually identified with *Frames* [13]. In the context of Image and Signal Processing, Frames are also named Schemata [23], as elementary concepts inside a network often refer to self-consistent chunks of knowledge necessary to reach a perceptual goal. A node of a Frame network becomes a complex data structure representing not only the declarative role of a concept but also its stereotyped functionality in the context of one or more problems. From the point of view of Computer Science, the Frame concept is strictly related to Object-Oriented languages as well as to the concept of agents inside DAI systems.

Semantic Networks could be either Extensional or Intensional Systems. We recall that Intensional systems [2] must use explicit representations of both the status of a system and of the related uncertainty; this allows them to identify a general mechanism of uncertainty propagation that does not depend on a specific problem but can be applied to a whole class of concepts. Using Semantic Networks makes it easy to attach a truth value to each instantiated concept, by allowing the corresponding prototypical concept to contain an appropriate uncertainty attribute. Consequently, the inference method used together with the SN has no difficulty, at least in principle, in manage two-directional inferences as well as in drawing conclusions about the uncertainty of a concept, on the basis of the knowledge coming from correlated sources. These are two of the main problems related to Extensional systems [2].

2.3 Uncertainty Management Techniques

Uncertainty management plays a key role in IMP systems, as well as in DFSs, for two main reasons.

First, noise due to physical phenomena and to discretization effects in analogic-to-digital conversion affects input data. Consequently, the observed representation of the world does not correspond exactly to the monitored situation but only approximates the actual situation. Secondly, knowledge allowing a system to associate input data with causes is often partial and incomplete. As a consequence, the solution to a problem, even in the presence of ideal input data, can be assessed only to a finite degree of precision.

Different techniques have been used inside KBS systems to describe uncertainty and to manage its propagation. A proposed classification [2] divides such techniques into probabilistic and non-probabilistic ones. Bayesian laws are used to draw conclusions in the first case. Fuzzy sets theory [7], and Dempster Shafer rules [8] are not probabilistic methods that have been used in the context of IMP systems.

As mentioned earlier, techniques also depend on the type of KBS system employed, either Extensional or Intensional. Uncertainty management techniques employed in Extensional systems are based on logic, as they refer uncertainty and the related propagation rules to ideal symbols. Each uncertainty value attached to a symbol depends only on the uncertainty value of the rule by which it is assessed and on the related values of pre-conditions. In Intensional systems, symbols must have an additional property, i.e., a subset of them must represent a self-consistent part of the solution space that can be identified with a (partial) status of the overall system. This means that uncertainty must refer to a class rather than to single concepts. Changing uncertainty related to an instantiated symbol by using a certain rule has automatic cross effects on all symbols belonging to the same subset, or linked to the symbol by causality relationships. This also means that a KBS Intensional system must also be able to perform non-monotone reasoning, e.g., to change its conclusions on the basis of new evidence. This is particularly useful in IMP systems, where backtracking of errors made in image processing and in recognition phases is often necessary.

Among the techniques used in Intensional systems, a probabilistic one should be mentioned for the great interest that has aroused in the vision community in these last few years [9]: Bayesian Belief Networks [2]. This is mainly due to the fact that this theory provides a inference mechanism that well explains both Top-Down and Bottom-Up aspects of propagation of information in the control flow of an IMP system.

3. Distributed Knowledge-Based Systems

A major issue of the research on AI and IMPs in recent years has been to assess the potential and practical benefits coming from a greater availability of parallel hardware. In vision, both coarse-grain and fine-grain parallelism has been exploited in order to search for efficient solutions to a complex problem.

On the other hand, people working on KBSs have realized that decomposition of the rules contained in a PS into separate sets, each pursuing a subgoal in the context of a complex problem [31], is a good strategy to develop more robust and more efficient systems. Analogies to the modular structure of certain complex functionalities of the brain have supported this point of view from a biological standpoint.

These aspects have convinced people that KBSs (in particular, IMPs) may be successful provided that one studies how knowledge can be distributed among multiple entities and that advantages can be obtained by using such distribution.

Further, following the ideas contained, for example, in the Frames representation theory, the necessity for allowing each separate chunk of procedural and declarative knowledge to be applied separately by parallel agents has been pointed out. This implies that a Distributed KBS can be represented as a set of N autonomous knowledge sources, and each one is characterized by a separate Knowledge Base KB_i, and an inference method M_i, and is able to process (in general) different sets of input data I_k to produce parts O_k of the output solution space (see figure 2).

At this point the foundations of DAI (Distributed Artificial Intelligence) were laid: it was clear that the study of distributed systems integrating autonomous and intelligent knowledge-sources implied the necessity for solving the problem of maintaining consistency among different parts of the solution space. Increasing consistency when the number of agents becomes large requires a high level of co-ordination and cooperation, which can be maintained mainly by increasing the amount of information circulating between the nodes of a system.

Both Hardware (i.e., multiprocessor) and Software (i.e., multiprocess) architecture's have been studied [11] since few years. Two main communication paradigms and possible combinations of them are currently investigated: shared memories and message passing.

In the first case, memories shared by knowledge sources are used as memory spaces to maintain common instantiated knowledge about the partial problem status. Semaphores or similar tools are used to regulate the access to such areas.

In certain approaches, also used in IMPs [11, 21], shared memories become blackboards, i.e., active system parts, able to signal which knowledge-

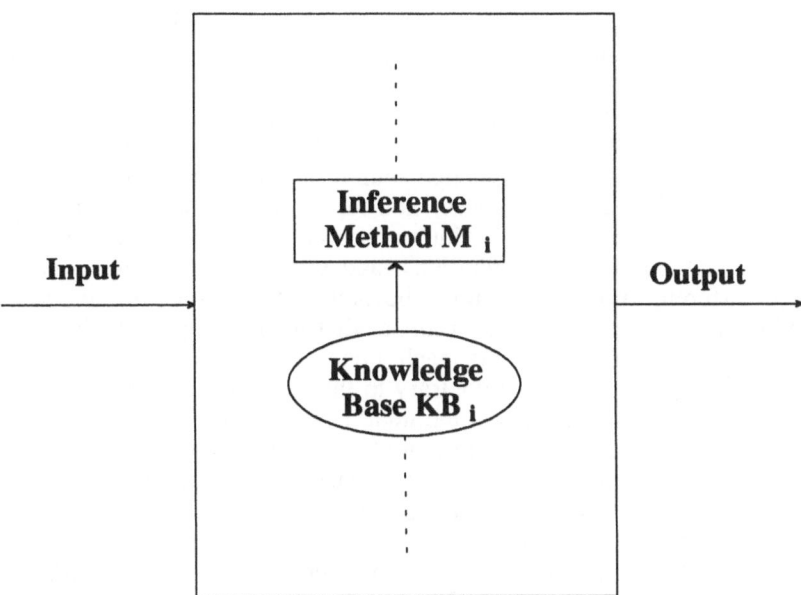

Fig. 2: Distribuited knowledge based system

sources (KSs) can provide their contributions, depending on the status of the solution space. However, blackboards have been proved to be a bottleneck for system performances due to the heavy communication load required at the centralized level.

A proposed solution for complex problems has been to split the global blackboard into a hierarchical network of local blackboards. This solution can be applied when KSs are organized according to functional hierarchy, too, as is often the case for IMPs. KSs operating at adjacent abstraction levels communicate through local blackboards, so that the possible number of communication channels to be managed and the amount of information to be transmitted by the agenda of the blackboard are reduced. If the number of local blackboards becomes comparable with the number of KSs this approach becomes similar to message passing.

Message passing provides a communication paradigm more abstract than shared memories. The behaviour of the agents in the net depends only by the content of the received message and by the node which sent the message. This approach has the main advantage of distributing the communication load by allowing multiple channels to be established among different KSs. However, it is more rigid and less flexible, in the sense that a new application requires to define the network structure as a part of a-priori knowledge about the considered problem. From the point of view of Software Engineering, this approach has a counterpart in Object Oriented programming.

The current trend of IMPs is to use Object Oriented programming by allowing local memories to be shared by linked modules, so splitting the concept of message from the concept of data. Messages allow the receiving module to select one among possible operations to be performed on data stored in the shared memory.

Once a communication paradigm has been chosen, the problem is choosing techniques able to maintain a global consistency among the conclusions drawn by each knowledge-source of a DAI system. In particular, due to imprecision, incompleteness, non-monotonicity of data and models uncertainty management and solution search techniques are necessary. Multidirectional inference contributions, and correlations among conclusions drawn by different knowledge sources become central issues of such systems. An additional problem arising from a distributed approach is to keep synchronization by also allowing each node of a DAI system to retract conclusions when new evidence contrary to the current status is available. The structure of the solution space and the relationships among partial results must be assessed, together with the definition at a global and at a local level of the process devoted to explore the search space. An important feature of a distributed system is that it is often possible (and sometimes necessary) to obtain a certain degree of redundancy (i.e., superposition of local solution spaces). This allows an increased degree of robustness to be obtained, but , at the same time, it opens problems related to the efficiency of the computational resources spent by each module for additional operations requiring communications with other nodes, such as comparison of results.

According to the above considerations, it appears evident that a DAI system must exhibit many properties that are provided by an Intensional approach to uncertainty management. Among Intensional approaches to DAI systems, one of the most theoretically sound is represented by Bayesian Belief networks. This probabilistic approach is an extension of Bayes theory to manage uncertainty inside networks of agents, each devoted to estimating a random variable. Links between nodes represent causal relationships, or, more generally, dependencies between variables associated with the linked nodes. A key hypothesis in Bayesian Networks is the possibility of defining a hierarchy inside the networks, i.e., the capability of establishing higher- and lower-level nodes for each node. This hierarchy corresponds to cause-effect relationships among variables of the net. For example, higher-level nodes may represent more abstract concepts that may cause the event associated with a node to occur, whereas lower-level nodes may be associated with consequences of the event itself. In this way, one can provide a probabilistic description of things and facts related to the considered problem.

Under suitable hypotheses, which can be verified by most IMPs, it is possible to show that for singly connected networks, there exists a distributed estimation technique able to fix the most probable value of each variable of a network in such a way as to maintain a global consistency, i.e., each estimated value is consistent with all others, in the sense that is part of configuration of values which maximizes a probabilistic criterion. Consequently, one can say that the solution space is defined by the set of variables estimated by the net, and that correlations between different parts of the solution space are described probabilistically. A frequent assumption in order to reduce the amount of knowledge (i.e., conditional probabilities) to be provided as a-priori knowledge to a system is that only variables in neighbouring nodes of the net depend each other (i.e., first order Markov independence is assumed inside the net). It is interesting to note how this concept has a close correspondence in the hierarchical communication paradigm of both message-passing and distributed blackboard. This is more clear if also the search mechanism is considered. Such mechanism is compound of two parts: first, each node is able to locally estimate its variables by maximizing a probabilistic measure, i.e., the belief, made up of two contributions: an expectation term coming from higher-level nodes and a evidence term coming from lower-level nodes.

Such terms are not directly available to a node, but they are computed by adjacent nodes and transmitted to it during the search process by a distributed message-passing scheme, which constitutes the part of the search mechanism devoted to maintain consistency among partial solutions. can be defined capable to make each node capable to transmit its expectations and evidence to adjacent agents. From the above discussion it is clear that Belief Networks have interesting properties to be considered as candidate tools for the realization of Intensional Distributed KBS. Here, it is sufficient to say that such a mechanism seems to be very powerful to represent IMPs. A target architecture could imply using Bayesian Networks together with an appropriate distributed set of agents. A further extension is to modify the local belief measure in order to allow each agent to locally provide its contribute.

Following [16], we could think to modify the belief measure by adding a third term which takes into account the local contribution of the agent together with expectations and evidence. In such way, an agent could be consistent, for example, of a Semantic Network representing specific knowledge of the agent and a general purpose PS, which produces new evidence, cooperates to the search by working on concepts such as belief, expectation, and evidence, and by managing local maximization and message passing strategies. Depending on the epistemological content of a network, it should be possible to develop two types of IMPs of this kind: object centered and process centered systems. In the following, such IMPs are considered in more detail.

4. Image Processing Applications

IMP applications of KBSs involve the representation and use of different kinds of knowledge. The knowledge about a scene and the objects contained in it and the knowledge about the information-processing structure constitute the most important bulk of the knowledge used inside IMPs. To better consider this point, the structure of DOORS (Distributed Object Oriented Recognition System) [14] is presented as an example of how a KBS can use both types of knowledge.

4.1 DOORS

The basic a-priori information source of DOORS comes from the consideration (common to all KBSs), that the knowledge about the objects that may appear in a given image is often available for a specific application. This knowledge usually refers to spatial-temporal scene models. Such spatial-temporal models can be represented as semantic networks at a very general level inside KBSs. In such cases, links between nodes assume different semantic values, depending on the spatial or temporal relations they represent [15].

For example, knowledge about the expected positions of objects, either with respect to an absolute point of view (e.g., a fixed camera) or to a relative one (e.g., the reference system associated with another object), are available for certain applications. Moreover, physical constraints may limit object movements, so allowing only a limited set of temporal relationships. In figure 3, an example of a semantic network for object representation is given.

As can be seen, descriptive knowledge about objects can be structured into a multilevel network, which can be viewed as a graph. The representation axis correspondent to the multilevel organization of the network aims at describing the expected relationships between data produced along the information-processing hierarchy adopted to solve a given IMP problem and the presence of particular objects in the observed scene.

Each network level describes spatial and temporal relationships among the descriptive primitives that can be associated with a certain set of objects at that level. At each level, the expected presence of an object can be represented by providing a graph whose nodes are descriptive primitives of the type extracted by the set of algorithms associated with the same level of the hierarchical structure of the IMP. Relationships between descriptive primitives represent different spatial and temporal relations which are expected to hold if an object

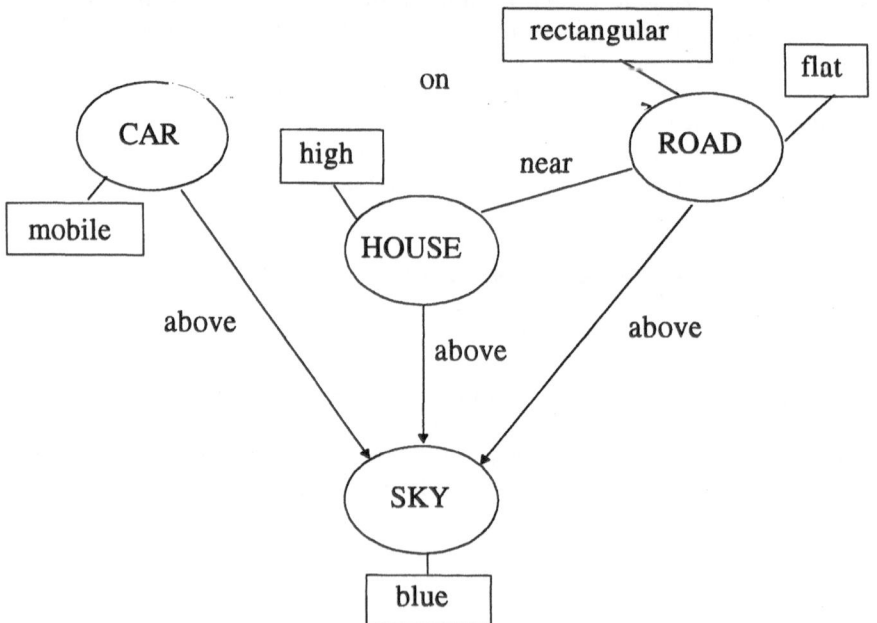

Fig. 3: A simple example of semantic network. Circles represent concepts, boxes represent attributes.

is observed: for example, the presence of a uniform object may induce a constraint on the possible luminance values of adjacent pixels, while a slow motion hypothesis implies similar luminance values to occur in close pixels belonging to consequent frames. The type of relationships among primitives on which constraints may be set depend on the abstraction level of the primitive along the net.

SLOT	TYPE	DESCRIPTION
Name	List	Name of the module
Higher-modules	List	Names of higher-level modules
Lower-modules	List	Names of lower-level modules
Higher-Requests	List	Top-Down Messages from higher-level modules
Urgent-Requests	List	Urgent Input Messages
Urgent-Answers	List	Urgent Output Messages
Lower-Data	List	Bottom-Up Messages from higher-level modules
Creates-Frame	List	Type of data produced by the module
First-TD-task	List	First set of rules to be activated when a top-down message is received
Interrupted Task	List	Name of the Task making the Urgent Request (useful for resuming after the urgent answer)
Do-Your-Job	Method	Local Inference Engine: activates sets of rules (tasks) of the module.

Fig. 4: The modules as the highest level knowledge source in DOORS: module slots are described and their type is specified.

For example, at the pixel level, the concept of adjacency can be immediately referred to the image lattice. At the edge level, parallelism, colinearity, and convergence relations between straight lines can be assessed.

From the above discussion, it is clear that DOORS is a distributed process-centered system, in the sense that the information-processing organization plays an active role in the overall system organization. Objects are used to parametrize the system functioning by providing different descriptions: searching for a different object may trigger implicitly particular behaviours and problem-solving strategies, depending at last from the available physical sensors and algorithms. For example, searching for a hot circular object may imply the selection of a description which may be provided only by an Infrared sensor: consequently, Top-Down activation of descriptions will select the description compound by a thermal image, a edge-extractor, and a circular Hough Transform to verify the object presence. As it can be seen, even though its process-centered nature, object descriptions are very important in determining the on-line functioning of the system, as they represent an associative key by which the distributed system may generate its behaviour. Therefore, it can be important to go a little deeper in the discussion about descriptive primitives organization.

Descriptive primitives differ from one another mainly in the types of causal relations that relate them to lower-level primitives. Two main types of relations are possible in an IMP: 1) a descriptive primitive can be referred to a different reference system with respect to a lower-level one. 2) a descriptive primitive can be caused by the presence of consistent groups (where the term *consistent* means verifying a certain grouping criterion) of lower-level primitives inside the same reference system. Combinations of the two types of relations are possible. For example, lines can be obtained by grouping edge [34, 35, 36, 37] points in the image space. Surfaces can be extracted by grouping lines into consistent sets in the image plane, thus forming, for example, well-shaped closed lines; then, such groups can be referred to another reference system, i.e., a 3D subject-centered reference system. Sometimes may also be necessary to refer objects to a general world reference system (i.e., a map), thus performing further coordinate transformations.

In DFSs [19, 32], multiple knowledge sources provide different observations about the same object. This implies the possibility of referring object descriptions to different kinds of descriptive primitives, i.e., by increasing the number of branches along the vertical direction of the Semantic Network. For example, see in figure 5 a graph related to the description of a car related to the use of a visual and a Infrared thermal camera.

To allow a IMP to identify a single cause of multiple possible perceptions of the same object, it is necessary to identify a third kind of descriptive primitives. Such fused primitives are required when more than one network branch provides (by using different algorithms) different primitives of the same kind (i.e., referred to the same object in the same coordinate system). Fusion consists of considering correspondences according to partial information, in order to estimate a single primitive representing a more complete, symbolic, integrated information of multisensory data. The implicit constraint is that a single description must result from multiple ones (unicity of the solution), or (considering the problem according from the point of view of object search) that a single cause must explain a complex effect obtained by fusing multiple observations. From these considerations, a definition for this third type of causal relation is the following: a descriptive primitive can be caused by the presence of a set of descriptive primitives each separately described in different reference systems. As one can see the key word of this definition are *separately*

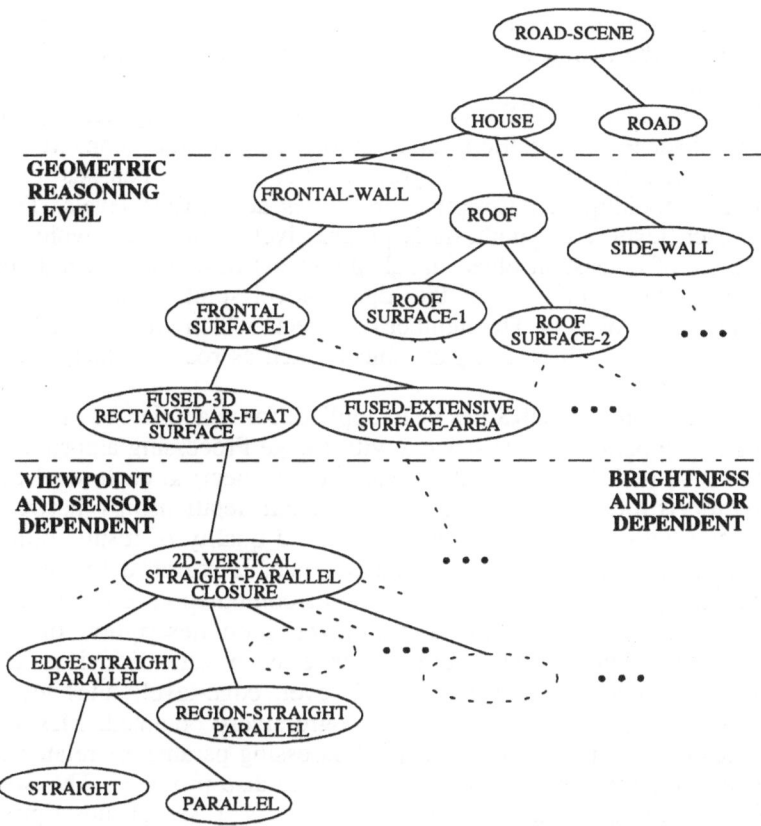

Fig. 5: Multilevel semantic network, partial description of a road-scene

and *different*; otherwise, grouping might be considered as a special case of fusion occurring locally among primitives described at the same description level. For example, closed sets of lines can be obtained by grouping single lines, or by extracting regions from an image and by identifying region contours. Fusion may be obtained by relating region boundaries and line groups and by creating a single description consistent with both.

In DOORS, knowledge sources associated with the highest level and with intermediate levels of the network are called virtual sensors, as they produce a virtual signal by performing a digital-into-digital transformation on the input representation.

Physical sensors (i.e., devices used to transform an analogic signal into a digital representation) can be regarded as the leaves of the network. In the case of IMPs, physical devices are devoted to acquiring images of an he observed environment from different positions. In this case, fusion occurs when descriptive primitives referred to a common reference system between the employed sensors are available. For example, if a room is monitored from two different viewpoints, a natural descriptive primitive on which to perform fusion is the position of a line in the room-centered reference system. If the process is performed correctly, the fusion output is a more robust and reliable descriptor of the environment.

In the first version of DOORS (written in LISP and C languages) [14], devoted to road-scene recognition and employed in the context of PROMETHEUS (an EUREKA CEC project), the solution consists in a context-dependent labelling of descriptive primitives at multiple levels. This means that, at each level, an hypothesis is progressively constituted which is compound of a set of consistent object label-(group of) descriptive primitives pairs. The global solution consists of the highest-level solution, i.e., the 3D description of the set of objects. The number of objects which are searched for is limited (i.e., less two or three complex objects such as road, obstacles, and signs are searched for).

In other applications, DOORS has been applied, for example, to estimate the best values of the parameters associated with Image Processing algorithms. In this case [16, 17] (CEC- MAST 0028c MOBIUS project) at each level the semantic network is oriented to describe functional relationships between produced data and between employed parameters and quality of results rather than to objects to be recognized. No particular object is searched for in this version; instead, the goal of the system is to obtain a combination of parameters which allows the system to provide a multilevel good description of the observed scene. Distributed quality criteria have been estabilished at the different levels (i.e., pixel-level, filtered-pixel level, edge-level, lines level) which allows the system to evaluate results progressively obtained. Message passing strategies taking into account effects of assessing parameters related to the behaviour of an algorithm at an adjacent level are also provided. The final result of this system, after an optimization cycle, is a set of descriptive primitives which correspond to the best quality object-independent description of the scene observed; such a description can be obtained by tuning the algorithm parameters. In this case, the solution space is compound by the parameters only.

A more recent application of the DOORS architecture has been realized in the context of the ESPRIT P5345 DIMUS project, devoted to exploiting the capabilities of DFSs for surveillance of underground stations. In this context, a C++ version of DOORS oriented toward real-time operations is inserted in a more complex system for crowding evaluation purposes [18,19]. Each level has a representation of possible alarms to be detected, each of which can be one-to-one mapped into a different range of values of number of the people in a station.

The distributed solution is represented by a belief measure for each possible crowding level; from this distribution, an estimated class of crowding may be obtained, by means of simple maximization, and it can be signalled upon request to the operator. In DIMUS, the solution space is structured on several descriptive levels: 1) rough digitalized primitives extracted from sensory data (i.e., images); 2) 2D descriptive primitives (circles, lines, profiles); 3) sensor based, fused descriptive primitives (cliques of virtual sensors); 4) isle-based, fused descriptive primitives (cliques of physical sensors); 5) station-based descriptive primitives (distribution of people in the station).

At lower levels, (from 1 to 3) the solution space is represented by the conditional probability of the estimated number of people in the sensor field of action given possible descriptive primitives at that level. At higher-levels, the probability of people is conditioned by the estimate provided by each sensor separately. A simple sum over multiple station isles is performed at the station level.

As it can be seen from the successive versions DOORS aims at being an Intensional system for IMPs, by exploiting different problems whose solution imply assessing possible partial configurations of a very complex search space. The divide-and-conquer strategy up to now followed allowed us to assess the applicability of such an approach, even though not all parts have still be implemented according to a pure Intensional paradigm. In DOORS belief networks theory is extensively used. The state space of the variables which represent the solution is explicitly defined, either it represents a association pair between an object and a descriptive primitive (recognition problem) or algorithm parameters (regulation) or an alarm situation (surveillance).

The system integrates expectations about most a-priori probable configurations in the state space and it collects evidences by mapping locally produced descriptions with possible local descriptions. In this way, each level may obtain either global or partial distributions of believes (depending on the search technique used). Uncertainty naturally arises from the relative score obtained by possible values of a variable, and the inference process has not to attach by means of rules an uncertainty value to each concept in the solution space. Even though these three cases have so far been considered separately, the long-term goal of DOORS is to embed in a single system an extended notion of status, which includes both recognition and regulation, thus enhancing the robustness of the resulting system, as suggested in Active Vision.

4.2 Other Systems

In DOORS a key role is played by the representation of the information-processing part. Active Knowledge Sources are associated with nodes of the semantic network that represent the different algorithms employed to progressively extract knowledge at higher abstraction levels. The semantic network used to describe objects plays only a parametric role with respect to the process network, as it provides different expectations for the descriptive primitives obtained. Other systems prefer to assign an active role to each object; the extreme case is to associate with each object inference methods and algorithms able to process an image in a way independent of other objects. This approach has the major drawback of being unsuitable from the point of view of the memory space required, as it implies high redundancies in object description (from the point of view of procedural knowledge). It is more advisable [30] that a limited number of copies of a general purpose processing mechanism should be present in an IMP (mainly for robustness purposes), which can be parametrized according to object properties. A blackboard approach based on an object centered representation can be maintained at a more symbolic level, for example to decide to which object to allocate one of the available   IMP resources. In the following, other existing systems are briefly described.

Prototypical Knowledge about objects can be well represented inside Semantic Network, thanks to their relational structure. This aspect was realized by the first developers of KBSs. Nagao and Matsuyama [20] proposed a KBS for interpreting aerial images: in that system knowledge sources were associated with objects to be searched for. Objects were classified as fundamental, secondary and optional, and searched for according to their related degree of importance in the observed scene.

The basic knowledge source of their system was the ODS (Object Detection Subsystem), which consisted of a set of intrinsic and relational descriptions of an object, together with a procedural strategy to decide the priority of actions to be performed during matching. Another classic KBS in Image Processing is the VISION [21] system, developed by following the ideas of HEARSAY [22], i.e., the first KBS for speech interpretation. In VISION, the concept of a visual descriptive primitive plays a key role, as it constitutes the basis for representing the knowledge about the data that must be extracted from an image to interpret it. Knowledge Sources associated with objects contribute to the interpretation of a scene by applying their own strategies to extract those primitives which are useful to establish the presence of an object in the scene. Such systems are blackboard systems, as they use an active shared memory to address, the knowledge sources that must be fired, depending on the status of the problem. Both Vision and Hearsay have a major drawback: object recognition is based on a set of features extracted from an image according to a fixed strategy.

This strategy either selects various algorithms or sets the thresholds to different values for the same algorithm, depending on the object to be searched for. However, it is not easy to represent different backtracking actions for different objects, especially for a large number of objects that may be present in a scene.

The Schema System [23] which has been developed starting from VISION exhibits the same drawback as the previous two systems, even though it tries to overcome the problems of a centralized blackboard by applying a distributed approach. Schemata are perceptual strategies (i.e., frames) to be applied in order to detect an object. Schemata are represented by attaching to each object the procedural knowledge required to search for it. Local blackboards are used in [23] to increase the efficiency of the system. The Schema System is also able to integrate information coming from multiple virtual sensors in the recognition phase, and to drive, to a limited extent, possible alternative acquisition strategies. However, multiple virtual sensors are only used to improve recognition robustness by providing different features, and not to develop a single, more robust representation.

A system for interpreting biomedical images, (i.e. 2D-slices or 3D complete sets of tomographic Nuclear Magnetic Resonance images) is the IBIS system [24], developed at DIBE in 1987. This system is based on Semantic Networks to describe anatomical knowledge and on a PS operating at the general level, by using object-independent recognition rules, which are utilized by a fixed processing strategies. The particular application domain of this system allows the use of well-assessed knowledge provided by experts (i.e., radiologists).

A more recent KBS in which object representation plays the main role is ERNEST [6], which particulary meets the requirement for providing knowledge to a recognition system at a sufficiently general level. The Semantic Network on which ERNEST is based has been shown to be epistemological adequate to representing the knowledge necessary to perform the spatial-temporal interpretations of a wide class of signals, such as those related to scintigraphic images, and industrial scenes, and even acoustic signals (i.e. speech). ERNEST is a distributed frame network in which frames are associated with objects or with their descriptive primitives.

A new trend of KBSs that is worth mentioning, is represented by the Intensional approach to scene interpretation as proposed by Modestino and Zhang [25], who use Markov random fields to model a-priori knowledge necessary to associate descriptive primitives with objects. A hierarchical approach to Image Understanding, based on Markov Random Fields to represent spatial-temporal knowledge at each level and on Bayesian Networks to represent inter-level relations has been also proposed by [26], even though they present only one example concerning lower levels (i.e., segmentation of pixels and regions). Probabilistic approaches to constraint representation have emerged, in the last few years, to solve almost all inverse problems involved in Image Processing. The issue of combining such numerical representations and the existing symbolic ones, to exploit their correlation and main features inside a single system, will be the greatest problem of future research on KBSs.

Some systems have tried to exploit the knowledge about the algorithms used inside a KBS and their functionalities is necessary. One of the first systems of this type is the PS for image segmentation designed by Nazif and Levine [28]. In this system, a centralized Production System is used to produce a reliable image segmentation.

There exist few systems that address the regulation problem in a KB way. Among them, the OCAPI system proposed by [29] is worth to be mentioned. This system is a PS that selects one from among multiple possible paths in a hierarchical chain of image-processing algorithms. In this sense, it differs from [10,17], where a fixed processing chain is selected, and the problem is to fix the parameters for the chain algorithms. A combination of both approaches may lead to a very robust IMP.

5. Conclusions

Knowledge-Based systems have always been considered as powerful tools for solving many image processing problems, as a-priori knowledge plays a basic role in reducing the complexity of the search phase in the solution space. This concept was recently assessed in a formal way by Regularization theory, and by other theories.

However, application of AI representation techniques to Image Processing has been considered naive by many authors in the last ten years. This has mainly been due to the consideration that existing knowledge about the visual process is usually too partial and incomplete to decide how to organize it in an efficient way. A bottom-up scientific approach to the problem was invoked by Marr [3] who originated the computational school, oriented toward the study of mathematically and physically motivated numerical techniques. In the last ten years, this approach has allowed a deeper insight into phenomena related to the processing of visual information. The growing interest in DFSs, which were first used few years ago [27], and which have drawn the attention also of the Vision community [33], seems to show that sufficient knowledge to solve several basic problems is now available, and that the major problem to be addressed is to integrate multiple knowledge into complex systems in an effective way.

As shown in this paper, Distributed Knowledge-Based systems has been used in this direction too, but only at the level of symbolic knowledge, yet reaching a certain level of understanding in general problem-solving approaches.

A close integration of such point of views and the development of systems able to take advantages from both specific and general techniques within the field of Image Processing and Understanding (i.e., applied methodologies able to address both different visual processes and their organization within a complex system) is the main goal of the next years.

Acknowledgments

The author wish to thank Vittorio Murino, Gian Luca Foresti, Max Peri, and Prof. Gianni Vernazza for enlighting discussions related to the content of this paper. The author gratefully acknowledge the technical improvement of this paper due to Angelo Milano.

References

1. Buchanan B.G., and Shortliffe E.H., "Rule-based Expert Systems, Addison Wesley", Reading, Massachusetts, 1984.
2. Pearl J., "Probabilistic Reasoning in Intelligent Systems: Networks of Plausible Inference", Morgan Kaufmann, San Mateo, California 1989.
3. Marr D., "Vision: A Computational Investigation into the Human Representation and Processing of Visual Information", Freeman, San Francisco, California, 1982.
4. Woods W., "What's in a Link Foundations for Semantic Networks", in Bobrow D. and Collins A. (eds.), Representation and Understanding, Academic Press, New York, 1975.
5. McDermott D., "The last Survey of Representation of Knowledge", in Proceedings AISB Conference, pp. 206-221, Hamburg, W. Germany, 1978.
6. Niemann H., Sagerer G., Schroder S., and Kummert F., "ERNEST: A Semantic Network System for Pattern Understanding", IEEE Trans. on Pattern Analysis and Mach. Intell., Vol. 12, No 9, pp. 883-905, Sept. 1990.

7. Zadeh L.A., "Fuzzy Sets and Systems, North-Holland", Amsterdam, 1983.
8. Shafer G., "A Mathematical Theory of Evidence", Princeton University Press, Princeton, 1976.
9. Rimey R.D., Brown C., "Where to look next using a Bayes Net: Incorporating Geometric Relations", Proceedings - 2nd ECCV '92, Lecture Notes in Computer Science, No 588, pp.815-819, Springer-Verlag, 1992.
10. Murino V., Peri M., and Regazzoni C.S., "A Distributed Algorithm for Adaptive Regulation of Image Processing Parameters", Proc. IEEE Conf. on Systems, Man and Cybernetics, Vol.1., pp.259-264, Charlottesville, USA, 1991.
11. Jaganatthan V., Dodhiwala R., and Baum L.S. (eds.), "Blackboard Architectures and Applications", Academic Press, San Diego, California, 1989.
12. Sitharama Iyengar S., Kashyap R.L., Rabinder N. Madan (eds.), "Special Section of IEEE Transactions on Systems Man and Cybernetics on Distributed Sensor Networks", Vol. 21, No 5, pp.1027-1230, 1991.
13. Minsky M., "A Framework for Representing Knowledge" in The Psychology of Computer Vision, pp.211-277, McGraw-Hill, New York, 1975.
14. Foresti G.L., Murino V., C.S.Regazzoni, and G.Vernazza, "Distributed Spatial Reasoning for Multisensory Image Interpretation", Special Issue of Signal Processing on Intelligent Systems for Image Understanding (in press) 1992.
15. Tsotsos J.K., "A Framework for Visual Motion Understanding", Ph. D. Dissertation, University of Toronto, Canada, 1980.
16. Murino V., Peri M., and Regazzoni C.S., "Distributed Belief Revision for Adaptive Image Processing Regulation", Proceedings - 2nd ECCV '92, Lecture Notes in Computer Science, No 588, pp.87-91, Springer-Verlag, 1992.
17. Murino V., Peri M., Regazzoni C.S., and Vernazza G., "A Distributed Probabilistic System for Adaptive Regulation of Image Processing Parameters", submitted to IEEE Transactions on Systems, Man, and Cybernetics, 1992.
18. Bisio G., Regazzoni C., and Vernazza G., "Spatial Fusion", Esprit P-5345 Dimus Report DIMUS/42/dv/0001-00/r/A1, 1992.
19. Ottonello C., Peri M., Regazzoni C., and Tesei A., "Integration of Multisensor Data for Crowding Evaluation", Proceedings IEEE Conference on Systems, Man, and Cybernetics, Chicago, 1992.
20. Nagao M., and Matsuyama T., "A Structural Analysis of Complex Aerial Photographs", Plenum Press, New York, 1980.
21. Hanson A., and Riseman E., "Vision: A Computer System for Interpreting Scenes", in Computer Vision System, Academic Press, pp.303-333, New York, 1978.
22. Lesser V.L., Fennel R.D., Erman L.D., Reddy D.R., "Organization of the Hearsay II Speech Understanding System", IEEE Transactions on Acoustics Speech and Signal Processing,Vol. 23, No 1, pp.11-24, 1975.
23. Draper B., Collins R., Brolio J., Hanson A., and Riseman E., "The Schema System", International Journal on Computer Vision, pp. 209-250, 1989.

24. Vernazza G.L., Serpico S.B., Dellepiane S.G., "A Knowledge-Based System for Biomedical Image Processing", IEEE Transactions on Circuit and Systems, Special Issue on Image Processing and Applications, pp.1399-1416, Nov.1987.
25. Modestino J., Zhang H., "A Markov Field Model for Image Interpretation", IEEE Transactions on Pattern Analysis and Machine Intelligence, Vol. 14, No 6, pp 606-615, 1992.
26. Arduini F., Regazzoni C.S., and Vernazza G., "Multilevel GMRF-based approach to Image Segmentation and Restoration", to be published in Signal Processing, Vol. 33, No 3, 1993.
27. Clark J., and Yuille A., "Data Fusion for Sensory Information Processing Systems", Kluwer, Norwell, Massachusetts, 1990.
28. Nazif A.M., and Levine M.D., "Low Level Image Segmentation: An Expert System", IEEE Transactions on Pattern Analysis and Machine Intelligence, Vol. 6, pp.555-577, 1984.
29. Thonnat M., Bijaoui A., "Knowledge-based Galaxy Classification Systems", in Knowledge-based systems in Astronomy, Heck A., Murthag F. (eds.), Lecture Notes in Physics, 329, Springer-Vrlag, 1989.
30. Bertero M., Poggio T., and Torre V., "Ill-posed Problems in Early Vision", AI Lab. Memo No 924, MIT, Cambridge, Massachusetts, 1988.
31. Tsotsos J., "Analyzing Vision at the Complexity Level", Behavioural Brain and Sciences, 1990.
32. Luo R., and Kay M., "Multisensor Integration and Fusion in Intelligent Systems", IEEE Transactions on Systems, Man and Cybernetics, Vol.SMC-19, No 5, pp. 901-931, 1989.
33. Aloimonos J., and Schulman D., "Integration of Visual Modules: An Extension of the Marr Paradigm", Academic Press, San Diego, California, 1989.
34. Canny J.F., "Finding Edges and Lines in Images", AI Lab. Report TR-720, MIT, Cambridge, Massachusetts, 1983.
35. Davies E.R., "Machine Vision, Academic Press", 1991.
36. Marr D., Hildreth E., "A Theory of Edge Detection", Proceedings Royal Society, 1979.
37. Foresti G.L., Regazzoni C.S., and Vernazza G., "Grouping rectilinear Edges by the labelled Hough Transform", (to be published in Computer Vision and Image Processing: Image Understanding), 1992.

Spatial Fusion of Multisensor Visual Information for Crowding Evaluation

M. Peri, C. Regazzoni, A. Tesei, and G. Vernazza
Department of Biophysical and Electronic Engineering
University of Genoa - Via all'Opera Pia 11A, I
16145 Genoa, Italy

Abstract

Evaluation of crowding in complex environments is a problem currently addressed in the context of surveillance systems both to detect potentially dangerous situations (overcrowding) and for statistical purposes related to activity planning.

The project DIMUS (Data Integration in Multisensor Systems, ESPRIT P 5345) aims to attain such goals. In this paper, attention is focused on probabilistic Knowledge-Based techniques for statistical evaluation of crowding. To this end, a set of visual sensors are placed in a monitored scene to have different views of the scene itself.

The multilevel architecture of the system is modelled as a Bayesian Belief Network (BBN) of message-passing nodes. Each node corresponds to a virtual distributed processor that is used to obtain a probabilistic value of the locally detected crowding level.

Laboratory results, simulating real-life conditions, show that a good crowding evaluation can be achieved by the proposed approach.

1. Introduction

In the field of surveillance systems for complex environments, such as underground stations, airports, etc., the problem of crowding evaluation is important for two main reasons.

First, the presence of a large number of people (overcrowding), as well as the detection of unattended people in forbidden areas may result in potentially dangerous situations that must be signalized to the operator of a surveillance system.

Secondly, it can be very useful to know (in a probabilistic sense) the number of people present in a monitored environment for statistical purposes related to activity planning (e. g., traffic planning to save resources).

In the first case, a very fast but not particularly accurate detection algorithm is needed; in the second case, accuracy is the most important requirement, although it is often in contrast with speed.

In this paper, attention is focused on probabilistic Knowledge-Based techniques used for statistical evaluation of crowding. Interpretation results (i.e., the estimated number of people) can be used both for statistical purposes and to confirm or disprove with higher accuracy possible dangerous overcrowding situations detected by faster alarm-detecting modules.

At the present, a system prototype has been applied and tested in laboratories simulating a real underground station. A demonstration-room map shows the positions of the optical sensors (see figure 1). The system provides a crowding estimation for each active physical sensor in the room, by means of a graphical User Interface (UI) [9] that shows the fields of action, using different colours for different estimated crowding classes. The crowding map can be updated in a distributed and asynchronous way, whenever the crowding class estimate for a sensor is available. The crowding evaluation system consists of a hierarchy of separate modules that process data at different abstraction levels; the interpretation process is performed according to a distributed philosophy [1][5]. Data are first acquired by the visual sensors, and then processed by different virtual sensors (i.e., nodes by which feature extraction algorithms can be applied) [4]. The basis for feature extraction is a focus-of-attention approach, by which only those rectangular areas occupied by people in an analyzed image are selected for further processing. Then, a set of features are extracted [13][14][15] from each occupied area image: straight vertical edges in the lower part of each occupied rectangle represent legs; circles and maxima in the 1D upper profile represent respectively the heads of people close to and far from a sensor (i.e., unoccluded and almost completely occluded).

Each virtual sensor is a module, that locally processes data to characterize people and to associate the feature values extracted from data with a probability distribution function (pdf) for the people in the locally monitored area. Modules are organized into a network according to a probabilistic Bayesian approach [2][3] used to represent and update the system and the local status variables (i.e., the evaluated number of people present in the whole scene or in some of its subareas).

Each module takes part in a top-down (TD) and a bottom-up (BU) information flow along the hierarchical architecture. In the TD flow, a module propagates requests for "focus of attention" to connected lower modules (i.e., requests for specific data processing or propagation of a possibly present a-priori hypothesis about crowding to lower levels of abstraction).

In the BU flow, processed data (i.e., evidence representation at a module's abstraction level) are sent to higher-level modules for further processing. On the basis of local inferences, each module can propagate crowding estimates and alarms to a global supervisor that interacts with the DIMUS user interface (UI) [9]. An object-oriented approach, based on data hiding modularity and a message-passing mechanism, has been chosen for the system implementation [6][7][8].

In section 2, the crowding evaluation system and its Low, Middle and High Levels processing aspects are described. In section 3, performances about both virtual sensor modelling and inference mechanism are assessed. Finally, in section 4, a set of possible strategies to furtherly improve results are proposed.

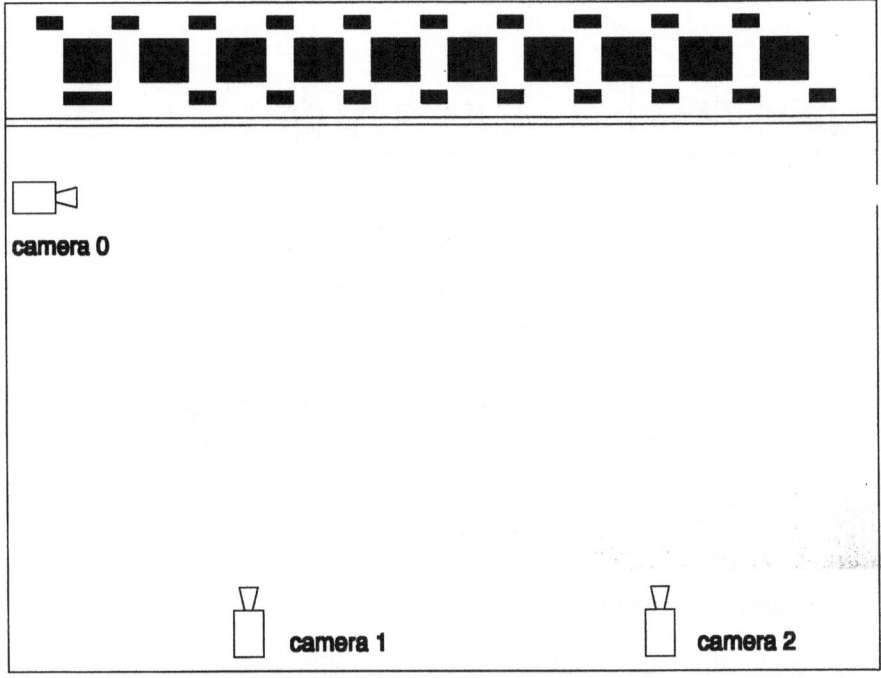

Fig. 1: Dimus intermediate demonstration room layout.

2. Description of a Knowledge-Based System (KBS) for crowding evaluation

Inside the multilevel architecture, inferences are possible by modelling the system as a probabilistic Bayesian network [3] of message-passing nodes. Each node corresponds to the instantiation of a virtual distributed processor that is used to obtain the probabilistic values of a stochastic discrete variable X, that is of the locally detected crowding level, and of the local status variable in the system. The representation of the status of each node X is completely defined by two probabilistic measures: the expectation π and the likelihood λ, which are estimated, updated, and propagated during the recognition process.

Whenever a node is activated by new evidences derived from data or by expectations from the operator, its belief value is updated according to the formula provided by Pearl [3]:

$$BEL(X) = \alpha\pi(X)\lambda(X),$$

where:

α is a normalising constant, $\pi(X) = Prob(X \mid e_X^+)$, and $\lambda(X) = Prob(e_X^- \mid X)$.

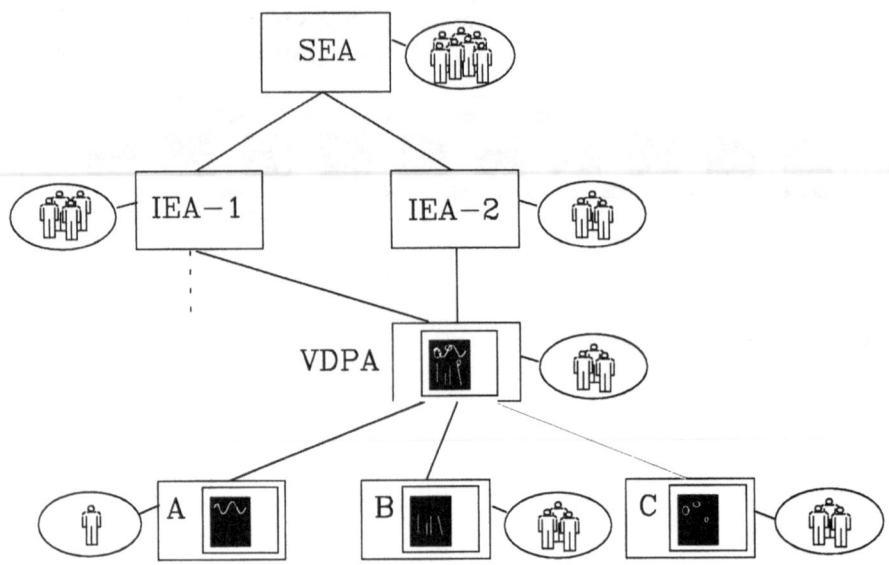

Fig. 2: Fragment of hierarchical inference architecture

2.1 The inference mechanism

In our application each node is associated with a discrete variable that represents the crowding level (see figure 2, 3) and that may assume one of few possible values. However, when we refer to the number of people, we deal with a variable that may assume a larger number of values. So we prefer to assume this variable to be continuous, and to use it for propagation of expectations and evidences among network nodes. On the other hand, we go on using a discrete crowding level locally at each node, to communicate alarm results to the human operator.

The system architecture provides a global status variable, that denotes the number of people detected in the whole station, and that coincides with the local variable at the highest-level node (Station Environment Analyzer). Moreover, the tree is logically divided into groups of one or more nodes which are devoted to the computation of the numbers of people present in certain subareas of the whole scene.

From the Bayesian network theory [2][3], one can deduce that the belief-updating process for continuous variables is deduced from that for discrete variables, and is well-founded only if three conditions are fulfilled:

- all interactions among variables are linear;
- uncertainty sources are normally distributed and are uncorrelated;
- the causal network is simply connected.

In our application, these conditions are satisfied, as:

- the stochastic variables are the numbers of people provided by different sensors located in the same or different subareas of the station, so they are linear combinations of such numbers;

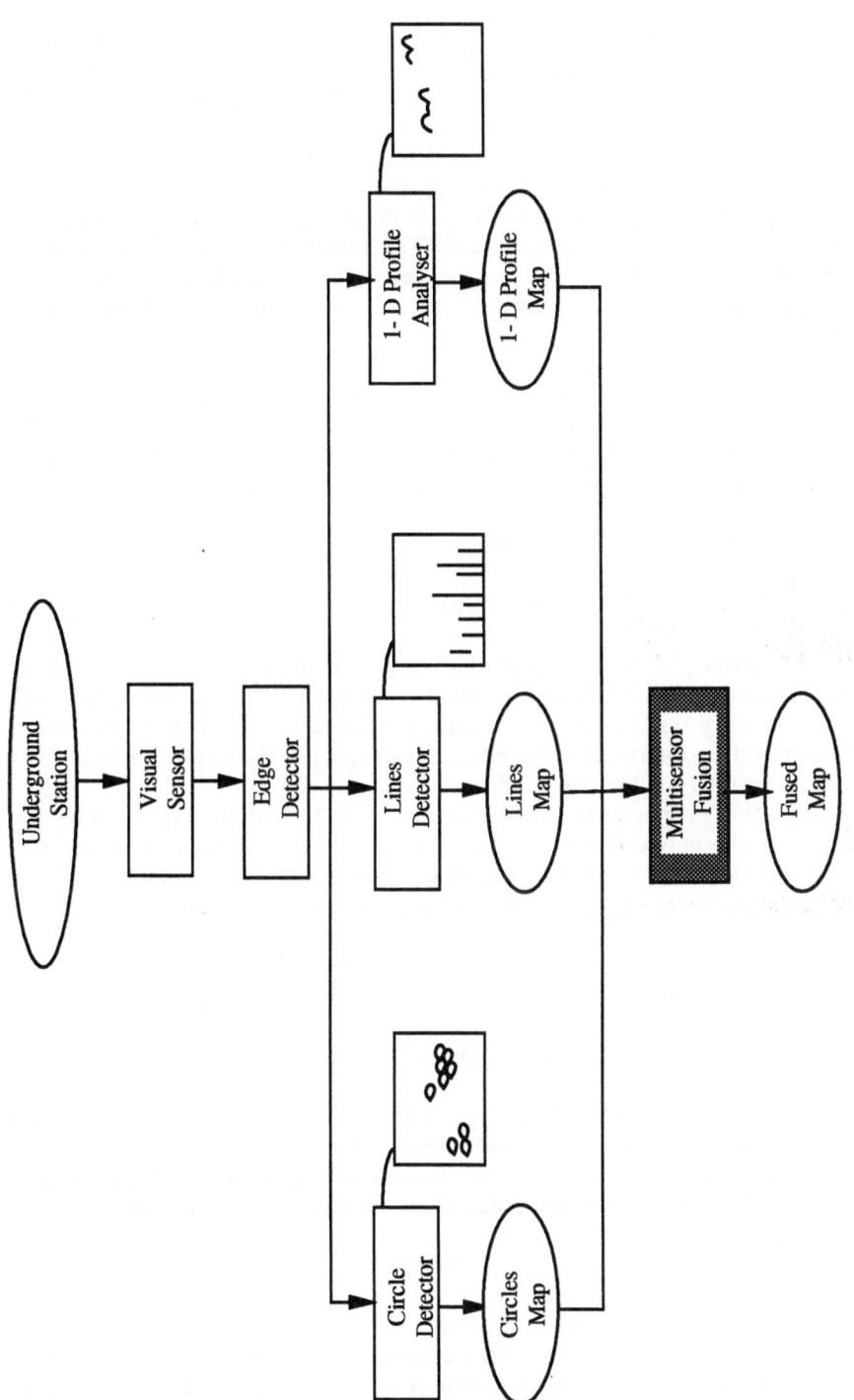

Fig. 3: Distributed fusion process

- virtual sensors (feature extractors) have been modelled as linear functions between the values of extracted features and the estimated numbers of people; and,on a statistical basis, a normal probability density for such features has been found; so also the stochastic variables for the numbers of people must have Gaussian distributions. In particular, from statistical considerations and a-priori reasoning, we have deduced that modelling normal density in an *asymmetric* way is more realistic. Unfortunately, this requires a higher computational load. The uncorrelation condition can be assumed: given a certain kind of virtual sensor, the probability of finding the corresponding feature is assumed to be uncorrelated with the other parameters;
- it is sufficient to look at our architecture scheme to realize that cycles occur in the highest part of the architecture. However, the chosen Bayesian approach uses only a one-shot TD-BU flow: once likelihood has risen up from the bottom to the highest node, the inference mechanism stops, without any further expectation propagation; this behaviour prevents any possibility of loops.

Finally, under this condition, we have:

$$\pi(x) = \pi_x(u)/k \tag{1}$$

(in the TD phase, it is sufficient that only one parent u should send its expectation to the child x). k is a constant characteristic of the son/father link that depends only on the respective controlled areas, under the hypothesis of uniform crowding distribution in those areas; $\pi_x(u)$ is the expectation sent by parent U to son X (it is the parent's local π).

About the propagation and updating of $\lambda(X)$, some explanations are necessary. In the network, we can distinguish two types of brother nodes, that is, virtual sensors monitoring a certain subarea of the whole environment at different levels. There are concurrent and complementary nodes.

Concurrent nodes take part in the estimation of the same variable x_1 (computed by their father nodes) by properly combining their own contributions x_{11}, x_{12}, $..x_{1n}$ according to the coefficients derived from a training phase:

$$x_1 = b_{11}x_{11} + .. + b_{1n}x_{1n}$$

In this case: $\lambda(X_1) = \Pi_i \lambda_{1i}/b_{1i}$. This case refers to virtual sensors monitoring the same area; they reason about different features extracted from data on that area. On the other hand, we can also deal with complementary variables x_i; the sum of these variables gives the value of another variable, y, at a higher level:

$$y = \sum_i x_i.$$

It is not a weighted sum, as x_i refers to different entities with respect to y. For example, each node dealing with a different x_i may be a sensor monitoring a different area of the station. So a global evaluation is obtained by summing of the numbers of people in the complementary and separate subareas. In this case, the probabilistic density for y is obtained by convolving and not multiplying the single x_i densities. As models are related to alarm situations, a belief measure indicates the degree of confidence for a situation occurring in the controlled environment.

2.2 Crowding modelling and feature extraction methods

For the specific purposes of crowding recognition and estimation, the object "crowd" is implicitly modelled as a collection of single persons by using a hierarchical modelling philosophy. A crowding level is described by a particular model or by a combination of subcrowding models (i.e., instances of crowding models belonging to the lower levels of the knowledge tree).

For example, in the present testing phase of the DIMUS project, ten people may constitute just one group or various subgroups made up of one up to nine persons (sub-crowding), and the single group made up of ten people may be a sub-crowding model of a group of fifteen, twenty, or more people.

Each sub-crowding area is associated with an occupancy rectangle on the image plane. The extraction of occupied areas allows focusing the recognition process only on the corresponding image parts (windowing), with a notable saving in computational resources. In order to perform as accurate as possible evaluations of the number of people in the station, each sub-crowding area is characterized by several hints, which are matched with the data obtained by using the related low- and middle-level algorithms:

- rectilinear edge extraction;
- 1-D profile extraction (associated with head profile);
- circle edge extraction;

Change detection and the focus-of-attention mechanism are the basis for feature extraction. Change detection algorithms produce a *foreground* image which points out those image subareas that present remarkable differences with respect to the background image. Two possible change-detection algorithms have been considered: change detection at the pixel level and at the edge level. The former operates by computing, for every pixel, the absolute values of the differences between the actual and the background images and by selecting foreground pixels by applying a hysteresis criterion (to the calculated differences) that takes into account the assignments of neighbouring pixels.

This algorithm, even though theoretically valid, is unsatisfactory for images in use because it is impossible to find a pair of thresholds that allow one to reasonably eliminate shadows of persons and, at the same time, to detect objects in poor contrast with the walls of the show-room. Therefore, change detection at the edge level has been chosen: the actual and the background images are considered after the edge extraction phase.

A Sobel edge detector [13] is used. Pixels with large gradient amplitudes (i.e., larger than a predefined threshold) are considered as edge points. Other edge-detection algorithms (in particular, the Marr and Canny [14] algorithms) have been tested, but the Sobel one has been chosen because it requires very short times (unlike Canny's), it seems more robust than the Marr operator, and it computes an estimate of the image gradient, which is very useful for further processing. The goal of focus-of-attention algorithms is to divide a foreground image into logically self-standing subareas. Our algorithm first searches for the minimum bounding rectangle that contains all foreground edges (i.e., all changes with respect to the background). Subsequently, the edges thinning and filtering phases follow, which eliminate noisy, very short and isolated edges. Finally, a reiterated threshold controlled splitting of the minimum bounding rectangle is performed so as to detect a minimum bounding rectangle for each change subarea. During the last phase, very small rectangles are eliminated.

The line-detection process has been implemented because of the importance of straight lines for the detection of the bodies (i.e., legs and arms) of nearly "unoccluded" people present in the environment (usually, people closer to the visual sensor). Line-detection is performed by using a Hough Transform-based algorithm [15]. The algorithm transforms the numerical information contained in a scene-discontinuity image into a relational graph, whose nodes are represented by segments, and whose arcs are represented by relational properties between nodes. It operates on the edge image and on the gradient-phase image.

Circle detection has been implemented, as circumferences are important to detect the heads of "partly occluded" people present in the underground station environment (usually, persons far from the sensor). The algorithm is based on the Circular Hough Transform method.

One-dimensional profile analysis has been implemented to define a first, alternative method for detecting and locating the heads of persons very far from the sensor (so integrating the circle detector). The algorithm operates on every minimum bounding rectangle that has been obtained by the previously described focus-of-attention procedures. First, a one-dimensional profile consisting of as many points as those of the rectangle width is calculated by considering, for every image column, the distance between the rectangle top-side and the first edge point. An interpolation phase is performed to "regularize" the profile.

To detect heads, a robust algorithm for the calculation of maxima has been implemented: for every maximum, it considers the local pattern in its neighbourhood in order to discriminate between a "lonely" maximum (isolated person) and a "composite" maximum (close persons).

2.3 Virtual sensor modelling

Modelling [10][11]consists in finding correlations between the values of each feature type extracted from an image by low-level algorithms and the number of people actually present in the same acquired image (see figure 4). A training set of images has been used to get the necessary measures related to the "feature values/people" plan and a linear interpolation of measurement points has been found to correlate these variables.

A linear interpolation of measures related to the XY plan (X = people, Y = value of the extracted feature) can be obtained by means of the Least. Squares Method, as follows:
- model: $y = mx + k$
- measures: $(x_0, y_0) (x_1, y_1) \, .. \, (x_n, y_n)$

- XP≅Y: linear system of 2 unknowns and n equations (for n>2, it is oversized), where:

$$Y= (y_1 .. y_n)^t \qquad\qquad P= (m\ k)^t$$

$$:X =\left[\begin{matrix} x_0.. & ..x_n \\ 1.. & ..1 \end{matrix}\right]^t$$

The presence of at least two different lines in matrix X (i.e., two measures with different x_i) is enough to ensure that P exists (as X has a full range):

$$P^*= (X^t X)^{-1} X^t Y.$$

It can be shown how far a linear approximation is from real measures by computing the curve of the estimated mean variance as a function of x. To demonstrate the different behaviours of this measure as a function of x, the lower and upper variances have been computed versus feature values given by a linear approximation.

Thanks to the variance-behaviour model, if the x value is found on the linear model corresponding to a given y value, it is possible to know with what uncertainty the number of extracted features is related to those people:

$$\text{left-sigma} = \sigma_{ly} \text{ and } \text{right-sigma} = \sigma_{ry}$$

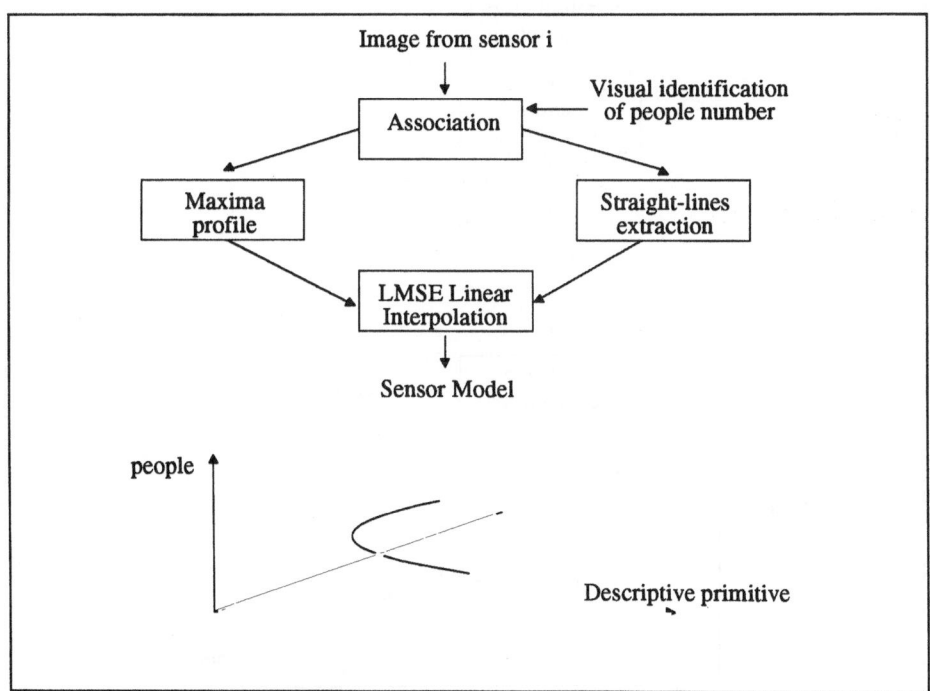

Fig. 4: Training phase

Because of the linear relation between feature values and people:
$$y = mx + k \Rightarrow x = cy + d, \text{ and } c = 1/m \quad d = -k/m.$$

Also the relation between their distribution densities is linear, in accordance with the basic theorem about distribution densities of dependent stochastic variables[4]. In this case:
$$(x = cy + d), \quad f_x(x) = (1/|c|) f_y(cy + d).$$

Consequently, also the densities related to people exhibit asymmetric normal forms, with the characteristic parameters
$$\sigma_{lx}() = (|1/m|)\sigma_{ly}() \qquad \sigma_{rx}() = (|1/m|)\sigma_{ry}() \qquad meanx = \mu_x = cy + d$$

Different virtual sensor models are used, if the intrinsic and extrinsic parameters of a physical sensor (camera) are changed; the parameters to which models may be more sensitive are position, angle of inclination, and focal length.

Thanks to the expressions for the pdf parameters of the people number of people as a function of the feature values, in the test phase it is possible to deduce the density distribution of the number of people that may be present in each occupied area ($\lambda(x)$), if the value of the features extracted from that area are given. This constitutes evidence from data (see figure 5).

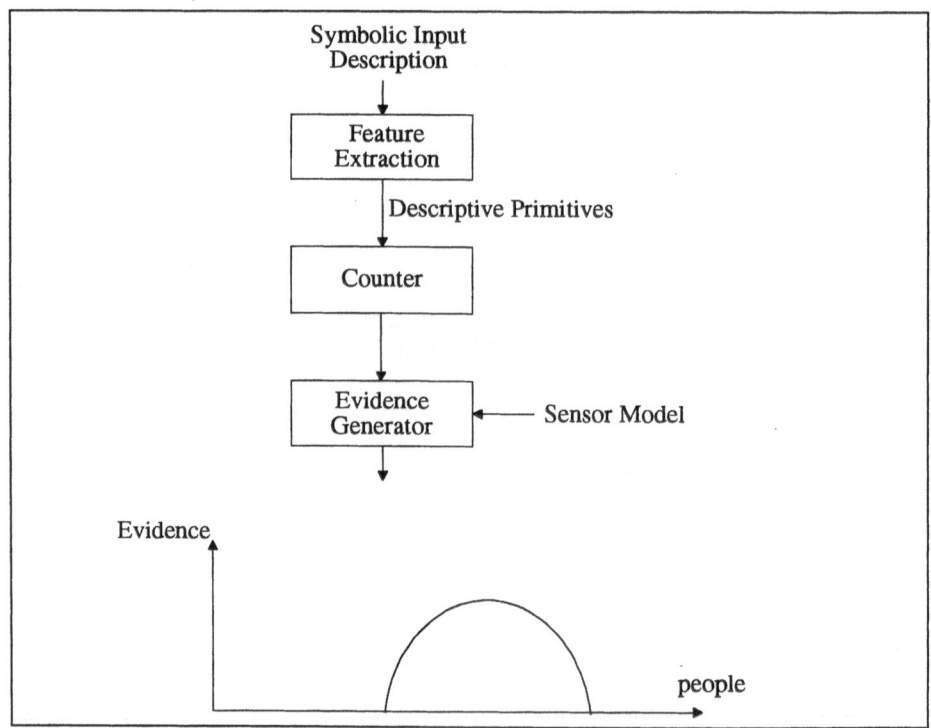

Fig. 5: Test phase

3. Results

Preliminary results [10] (obtained by tests performed on sets of about 200 images per sensor) show that an accurate crowding evaluation can be achieved by using the proposed modelling and inference approach.

About the system time required, the Low- and Middle-level phases take most of the whole processing time and their computational loads depend on the complexity of the scene analyzed (i.e., number of the present people). In the worst case, these phases require 15 seconds of CPU time per sensor. The High-level phase is faster; its computational load is independent of the number of people actually present in the image, and it takes less than 2 seconds of CPU time in a SUN Spark-Station2 machine.

In this section, only the Bottom-Up process is described, that is, no a-priori knowledge is provided to the network in order to avoid the positive or negative influence of external suggestions and to allow an actual testing of
- the virtual sensor modelling;
- the propagation and fusion mechanisms.

Modelling has been performed by means of training sets consisting of 300 images per sensor.

From the training measurements and intuitive considerations, it has been possible to deduce that a linear interpolation is a good choice only for some portions of the total measurements ranges. The best performances were obtained for small numbers of people because people do not hide one another and it is easier to assign to each of them one maximum belonging to the 1D upper profile, one circle for the head, and from 2 to 4 vertical edges for the legs. As the crowding level increases, much more occlusions occur; so, even though the curve remains quite linear, it shows a certain decrement. In the overcrowding case, a saturation phenomenon occurs: if a certain threshold is exceeded, the number of people becomes so large and occlusions are so frequent that the number of extracted features is quite constant. It is not possible to consider this kind of effect by using a linear model.

The choice of the most meaningful feature types can be made by considering the behaviour of the corresponding linear model (see Tables 1, 2): the closer to zero the angular coefficient (flat straight line), the more unreliable the model .

In order of decreasing significance, the features so far used are:
- vertical edges in the lower part of each area;
- maxima in the upper 1-D profile of the occupied region;
- circular edges.

It is possible to evaluate the crowding estimation performances at different levels of the hierarchical tree. The inference network considered in the paper is made up of nodes placed at five different levels (see figure 2). At the highest abstraction level, the Station Environment Analyzer (SEA) provides an evaluation of the station crowding level. At a lower level, each Isle Environment Analyzer (IEA) provides an evaluation of the crowding level detected in the corresponding station subarea (an isle is defined as the intersection between one or more sensor fields of action and the station platform).

	angular coeff.	origin ordinate
maxima in 1D profile	0.3	1.6
vertical edges	0.7	6.
circular edges	0.3	1.

Tab. 1: Linear model parameters for sensor 0

	angular coeff.	origin ordinate
maxima in 1D profile	0.44	1.04
vertical edges	1.66	6.5
circular edges	0.91	1.66

Tab. 2: Linear model parameters for sensor 1

Finally, the information coming from each virtual sensor at the lowest level (A: maxima, B: vertical edges, C: circles) is fused by the Virtual Descriptive Primitive Analyzers (VDPAs) to provide a crowding evaluation for each physical sensor.

Crowding estimation performances are evaluated in terms of deviation from the ideal correct behaviour, as the number of people varies. At present, crowding-estimation statistical errors and reliability values are given for the field of action of each sensor.

On average, it is possible to deduce that the statistical crowding estimation error at the sensor level is about 10%.

Tables 3, 4 give the system accuracy in crowding estimation for each sensor, with reference to the people in the related scene. The unreliability of sensor 0 in estimating many people is evident. For the other cameras, it is only possible to stress that the error increases as the number of people increases, too.

The increase in the statistical error as a function of the number of people is consistent with the appearance of occlusions and with the consequent saturation phenomenon above a certain threshold. Therefore, the estimation error might be reduced if a stepwise linear model were employed.

The following is an estimation example (see figures 6, 7). The background image and the current image (showing an overcrowding situation) are acquired by an optical sensor. These two images are used to extract the values of the three feature types (maxima, vertical edges and circles) in the occupancy area.

At the basis of the numerical value of each feature type, the corresponding local virtual-sensor model allows to associate with the rectangle a certain Gaussian pdf of the number of people; the value corresponding to the Gaussian maximum value is the estimated number of people.

After the inference process through the Bayesian network up to the physical sensor level, the system estimates the presence of "too many people" (a number larger than 15) with high probability (80%) in the corresponding sensor field of action. It is possible to note how much the estimation certainty increases from the feature-extraction level to the physical sensor level, where the fusion mechanism is performed.

Fig. 6: Background and current image

Fig. 7: Extraction of edges, maxima, vertical edges, circles from figure 6

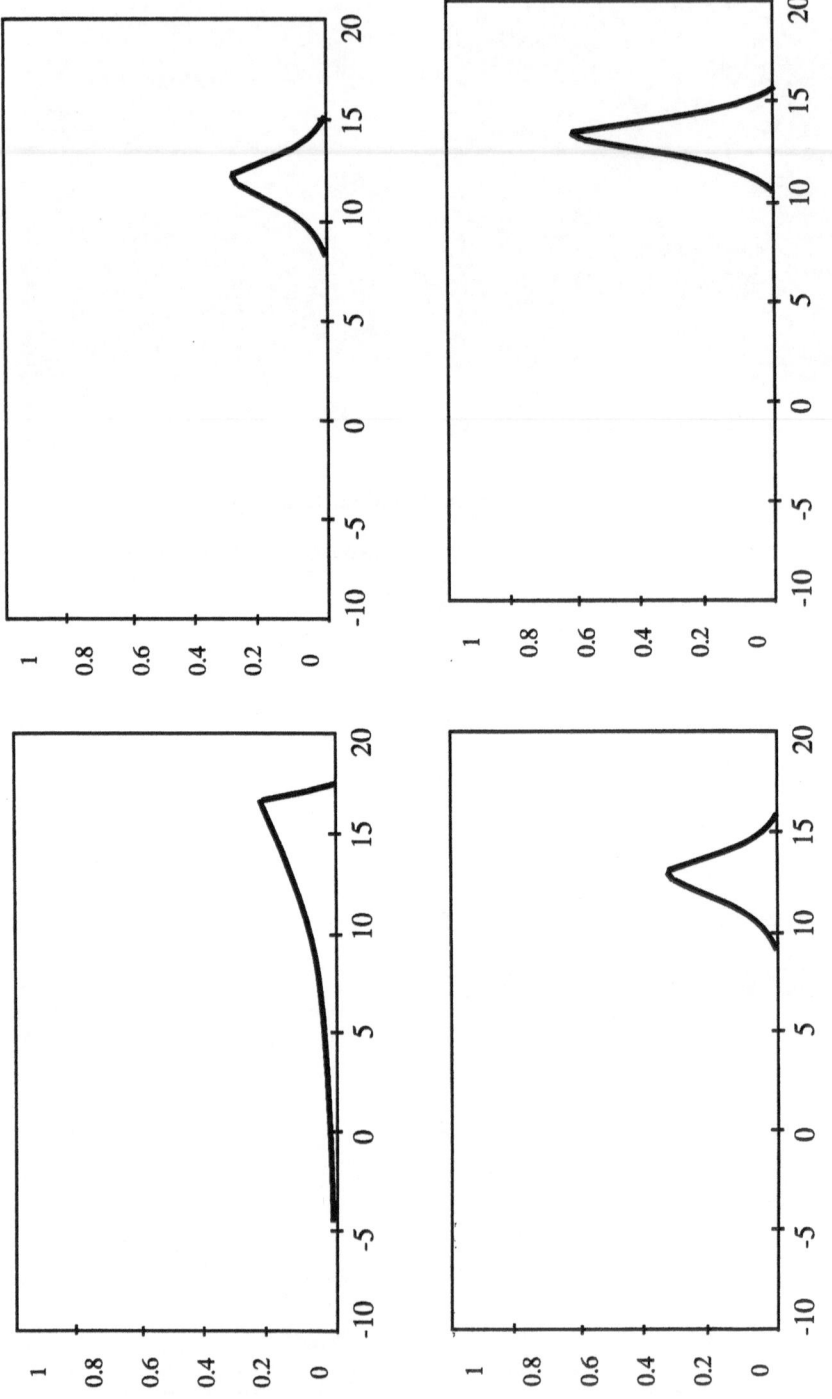

Fig. 8: Evidence pdf about the number of people

4. Future work

The system presents several limitations which will be overcome by means of a series of possible solutions. The network has been implemented for the final purpose of estimating the number of people present in the whole station or in each intersection of the sensor fields of action (isle). At present, the evaluation is restricted to the field of action of each camera separately. So, the Bottom-Up flow has to be completed from the sensors to the station, through the isles.

	1	2	3	4	5	6	7	8	9	10	11	12	13	14	15	16	17	18	19	20	21	22
1	2																					
2	3	2																				
3		2	3	1				1														
4		2		1																		
5			1	1																		
6				1	1		4		3	3		1				1						
7		2					3			4	1	3										
8							1	3	5	5	3	1	4	5		4		2				
9				4				2	2	4		4		3								
10								1		6		1		3		3						
11											3				1	2						
12												2	2									
13														1	2							
14													1		1							
15									1					2	2							
16														1	1		3	1	1			
17														3	2		1		3	3		
18															3			1	1			
19																				2		
20																	1		3	2		
21																2	2					
22																			1	2		
23																			1	2		

Tab. 3: Crowding estimation of sensor 0. Position (x=10, y=8) indicates that in 5 images the algoritm detected 8 persons from 10 present persons.

To improve the system performances, several techniques will be tested to speed up crowding estimation process and to optimize the robustness and accuracy of crowding estimates. Concerning the first objective, studies are under way to check on the validity of new and simpler features. This study phase might lead to hardware implementations of the extraction algorithms. Another improvement might be obtained, if the system could be implemented on a multiprocessor machine. In the present network parallelism is simply simulated. About estimation accuracy, current work aims to refine the linear modelling approach by means of a one-mode function with a linear behaviour at intervals. By using this new approach, problems posed by occlusion phenomena and saturation effects should be overcome.

This study phase might lead to hardware implementations of the extraction algorithms. Another improvement might be obtained, if the system could be implemented on a multiprocessor machine. In the present network parallelism is simply simulated. About estimation accuracy, current work aims to refine the linear modelling approach by means of a one-mode function with a linear behaviour at intervals. By using this new approach, problems posed by occlusion phenomena and saturation effects should be overcome. Higher robustness and precision in crowding estimation should be reached by using several models for each virtual sensor. Each model should associate extracted feature values with estimated numbers of people on the basis of different environment conditions(e.g., knowledge of light intensity, time etc.).

Finally, it is worth noting that the present estimation approach has no memory in terms of time; actually, temporal estimation (e.g. the application of Kalman filtering and prediction phases [12][16]) could be useful to predict crowding values (without having new data to process) between two successive estimates, both derived from data processing.

	1	2	3	4	5	6	7	8	9	10	11	12	13	14	15	16	17	18	19	20	21	22
1	10																					
2		10		1																		
3			6	3																		
4			3	4	2																	
5						1	3															
6						2	1	2														
7							2	6	4	2	2	1										
8								2	2		4											
9							1		4	2	7	4	2									
10											3	1	1									
11											1	1	1									
12												1	1	4								
13											3	6	6	4	3	1						
14											1	2	4	6	3	5						
15														2	5	1						
16														2	4							
17														3	2							

Tab. 4: Crowding estimation of sensor 1. Position (x=10, y=8) indicates that in 4 images the algoritm detected 8 persons from 10 present persons.

Acknowledgements

The authors wish to thank all the partners of the DIMUS project for their valuable cooperation on the system development. They are grateful to Paolo

Moretti and Paolo Delucca for their assistance in the Low- and Middle-Level algorithm implementation and tests.

References

1. G. Capocaccia , A. Damasio, C. S. Regazzoni, and G. Vernazza, "Dynamic Evaluation of multiple Sensors for Obstacle Detection and Identification", in Time-Varying Image Processing and Moving Object Recognition, vol. 2, Proceedings of the 3rd International Workshop, Florence, Italy, May 29th-31st 1989, Elsevier, 1990.
2. J. Pearl, "Probabilistic Reasoning in Intelligent Systems: Networks of Plausible Inference", Morgan-Kaufmann, San Mateo, CA, 1988.
3. J. Pearl, "Uncertainty Management in Expert System", Tutorial IJCAI'87, Milano, Aug. 1987.
4. R. Luo, M. G. Kay, "Multisensor Integration and Fusion in Intelligent Systems", IEEE Trans. on Systems, Man and Cybernetics, Vol. 19, No. 5, pp. 901-931, Sept./Oct 1989.
5. G. L. Foresti, V. Murino, C. S. Regazzoni and G. Vernazza, "Distributed Spatial Reasoning for Multisensory Image Interpretation", Signal Processing (in press).
6. B. Cox, "Programmazione Orientata agli Oggetti - Teoria, Tecniche e Sviluppo", Addison-Wesley Editoriale Italia, 1990.
7. B. Cox, "Message/Object Programming: an evolutionary Change in Programming Technology", IEEE Software, pp. 50-61, Jan. 1984.
8. D. Halbert, P. O'Brien, "Using Types and Inheritance in Object Oriented Programming", IEEE Software, pp. 71-79, Sept. 1987.
9. M. Ferrettino, A. Bozzoli, "A Surveillance System Project", ESPRIT Day ECCV, 1992.
10. P. Antognetti, A. De Gloria, P. Delucca, G. Vernazza, A. Tesei, "DIBE Activities after Intermediate Demo", DIMUS Project Internal Report, Jan. 1992.
11. C. Ottonello, M. Peri, C. Regazzoni, A. Tesei, "Integration of Multisensor Data for Crowding Evaluation", Proc. of IEEE International Conference on SMC, Chicago, pp. 791-796, Oct. 1992.
12. R. E. Kalman, "A New Approach to Linear Filtering and Prediction Problems", Trans. ASME, Series D, J. Basic Eng., pp. 35-45, 1960.
13. E. R. Davies, "Machine Vision", Academic Press, 1990.
14. J. Canny, "A Computational Approach to Edge Detection", IEEE Transactions on Pattern Analysis and Machine Intelligence, Vol. 8, pp. 679-698, Nov. 1986.
15. G. Foresti and C. S. Regazzoni, "Hough-based Recognition of Complex 3D Road Scenes", 1991 SPIE Conference on Advances in Intelligent Robotic Systems, Boston, MA, USA, Nov. 1991.
16. S. Bittanti, "Teoria della Predizione e del Filtraggio", Pitagora Editrice Bologna, 1991.

Robust Multisensor Fusion
in Underground Stations

S. Pfleger, A. Milano
Technische Universität München
München, Germany

Abstract

A major problem of data fusion in multisensor systems is the evaluation of multisensorial data correctness in order to ensure a correct and complete detection of the event occurrence (i.e. false alarms and missing alarm problems). One method is to construct high-level integrated sensorial data; this is *bottom-up* processing. Another method is to make predictions from models that impose constraints upon the event detection in a complex scene, like an underground station; this is *top-down* processing.

This paper investigates the *bottom-up* approach in event detection based on a stepwise integration of the multisensorial data in DIMUS (Data Integration in Multisensor Systems, ESPRIT project 5345). Robust fusion of the sensor observations in the presence of sensor failures is ensured by redundant and diverse sensors. The current sensor reliability and a weighted voting decision strategy are used for the selection of the correct sensorial data. An efficient technique for evaluating the current reliability of the sensor observations during execution is also presented together with the construction of high level logical sensors for the detection of objects and persons in the prohibited areas of an underground station. Finally, the performance of the alarm reliability is investigated, and a Measure of Belief is defined based on statistical and dynamical alarm reliability indicators.

1. Introduction

Correctness of environmental observations and the diversity of the multisensor data are important factors which determine the quality of the event detection process in an underground station. Problems like false alarms and missing alarms must be expected and appropriate handling mechanisms are needed for masking the faulty sensorial observations. Redundancy and diversity are the main concepts which ensure correct alarm detection in the presence of faulty sensorial data. Two processing methods could be used. One method is to

173

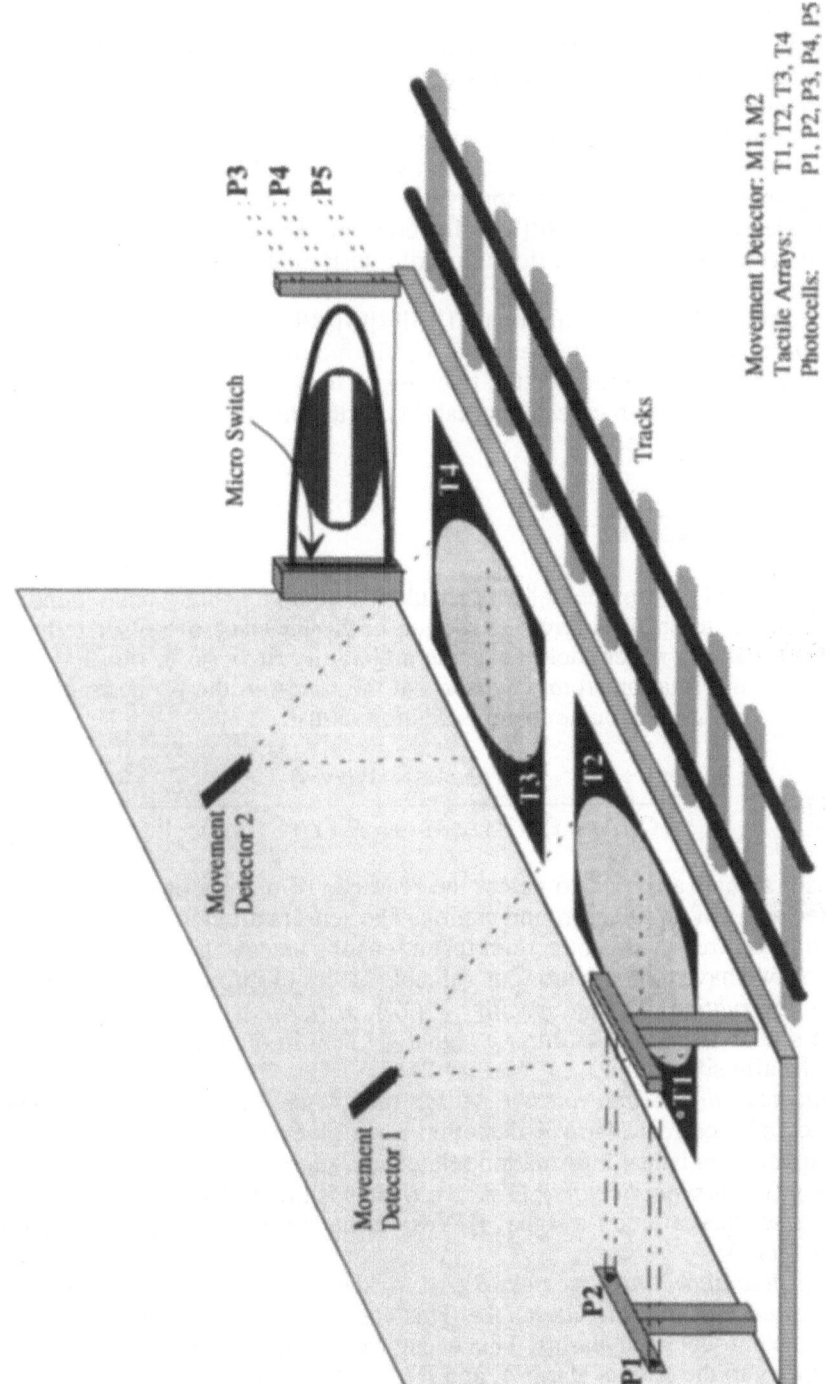

P3
P4
P5

Micro Switch

T4

T3

T2

Tracks

T1

Movement
Detector 2

Movement
Detector 1

P2

P1

Movement Detector: M1, M2
Tactile Arrays: T1, T2, T3, T4
Photocells: P1, P2, P3, P4, P5

Fig. 1: Topological Structure

construct high-level integrated sensorial data, i.e. to perform *bottom-up* processing. Another method is to make predictions from knowledge-based station models that impose constraints upon the event detection and interpretation in an underground station; here *top-down* processing is used.

An example for the first system type might be a *Multisensor System* (see figure 1), which could be attached to an existing visual monitoring system in order to provide complementary observations (i.e. tactile, movement data and beam breaker information [2][6][7]); this concept can be used in order to improve the robustness of the traffic monitoring task in an underground transport system.

An example for the second type [1][3][4][5][8][9] is a *Multimedia Tracking System* that could be attached to a traffic monitoring system in order to provide appropriate facilities for the tracking of dangerous situations in an underground station. In both cases the reliability of the multisensor alarm information plays a decisive role.

2. Alarm Reliability

The acquisition of environmental data should use complementary and redundant sensors in order to ensure correct execution in the presence of sensor failures. The alarm reliability is here defined as the probability *R(alarm, s, t)* of having a correct alarm information from a sensor *s* at the time *t* in the presence of both abnormal execution environment and execution faults :

$$\mathbf{R}\,(alarm,\,s,\,t) \;=\; \frac{P\,(Y/A,\,s,\,t)\cdot P\,(A,\,s)}{P\,(Y/A,\,s,\,t)\cdot P\,(A,\,s) \;+\; P\,(Y/A\,*,\,s,\,t)\cdot P\,(A*,\,s)} \tag{1}$$

The sensor signals are used to detect the presence of objects or persons in the prohibited areas of an underground station. The sensors are not "perfect" observers, and therefore we are not certain that a sensor *s* detects the presence of persons. What we have to go on are four valuable pieces of information:

- *the observation channel specifier P (Y/A, s, t)*: this is the probability that the sensor *s* sends at a time *t* a signal $\underline{Y}$ (Yes) in the presence of a specified alarm situation *A*.

- *statistical alarm correctness P (A, s):* the known probability of the occurrence of a correct alarm *A* detection for a given sensor type *s*. This is a long-term statistical information related to a sensor type.

- *the false alarm indicator P (Y/A*, s, t):* this is the probability that the sensor *s* sends at a time *t* a signal $\underline{Y}$ (Yes) in the absence of a specified alarm situation *A*.

- *statistical alarm incorrectness P (A*, s):* the statistical value related to the occurrence of missing alarms *A**. This value is sensor-specific and indicates the non-detected alarms. This event, called "missing alarm", is complementary to the correct alarm *A*, and *P (A*, s) = 1 - P (A, s)*.

The parameter *P (A, s)* denotes the probability of having a correct alarm based on the long-term statistical data related to the failures of this sensor type (e.g.

$P(A, s) = 0.999$ for each infrared movement detectors of the type VISOLUX PIR 30/32). $P(A, *s)$ is the statistical value of alarm incorrectness (e.g. $P(A*, s) = 0.001$ for the sensor of the type VISOLUX PIR 30/32).

The *alarm availability* is the most important quality factor. It is expresed in terms of the probability of having a correct alarm signal *at any* given time t. Highly reliable sensorial alarms, and additional dynamic fault masking mechanisms are needed in order to obtain a high alarm availability, and respectively low alarm non-availability. We denote by:

$$\mathbf{Q}\ (\text{alarm, s,t}) = [\ 1- \mathbf{R}\ (\text{alarm, s, t})\ ] \tag{2}$$

the dynamically calculated *non-reliability* of the raised alarm by a given sensor s at a time t. The occurrence of false alarms and missing alarms are the reasons why the alarm reliability of each individual sensor and logical sensor should be measured during their operation, i.e. dynamic alarm reliability evaluation.

We suggest to measure dynamically the false alarms of each individual sensors and to calculate its expected false alarm reliability which is measured based on a Scoring Scheme (see next section). The number of faulty alarms of each sensor is given by the *score* parameter.

A more complete view of the reliability of the sensorial alarm is here expressed by the *Measure of Belief* of the sensorial alarm at a time t. This is a tuple of five diverse reliability indicators:
1. the dynamic reliability of the alarm raised by this sensor at time t,
2. the observation channel specifier of the operating sensor at time t,
3. the false alarm indicator of the operating sensor at time t,
4. the statistical alarm correctness of this type of sensor, and
5. the number of already signalised faulty alarms by this sensor
(i.e. the score at time t).

The temporal behaviour of the alarm reliability of each individual sensor is described by the "*false alarm*" probability $P(Y/A*, s, t)$ together with its score value, and the observation channel specifier $P(Y/A, s, t)$.

3. Measurement of Sensor Alarm Reliability

The reliability of an operating sensor is measured during system execution. The comparison of the sensorial data obtained from the redundant and diverse sensors is used in order to detect faulty sensor information at a given time t.

We distinguish between normal and abnormal alarm detection as follows:
1. *correct alarm*: an alarm was raised in a specified alarm situation,
2. *false alarm*: an alarm was raised without any reason, and
3. *missing alarm:* an alarm was not raised in a specified alarm situation.

All reliability values of the sensors raising the correct alarm are maintained. The value of the "*false alarm*" indicator of the sensors which provided a false alarm signal is increased (e.g. by a value *weight*= 0.001) based on the selected scoring scheme. The score of these sensors is increased after each false alarm. Their alarm reliability decreases and their level of confidence, expressed by their *Measure of Belief*, decreases.

The alarm reliability of the sensors which provided successive false alarms decreases dramatically, and as a consequence their level of confidence becomes very low. The cumulative number of false alarms, is called *score*.

We specify a *Measure of Belief* **MOB** *(s, t_i)* which is attached to each sensorial observation at a given time t_i as follows:

$$MOB \ (s, t_i) := \{ \ R(alarm, s, t_i), \ P(Y/A, s, t_i), P(Y/A^*, s, t_i), P(A,s), score(t_i)\}$$

(3)

The dynamic evaluation of the new "false alarm indicator" is here performed periodically for each physical and virtual sensor, and this new value is used for updating a central table, called "False Alarm Scoring Table". Successive false alarms of a sensor will drastically decrease the reliability of its raised alarms, and will be followed by dynamic reconfiguration based on the replacement of the faulty sensor by a correct stand-by one.

False Alarm Scoring Example: Let us assume following initial values (at the starting point *to*) of the infrared movement detector sensor MD1 (see figure 1) :

R (alarm, MD1, to) = 1
P (Y/A, MD1, to) = 0.99
P (Y/A*, MD1, to) = 0.1
P (A, MD1) = 0.98
score = 0, and weight=0.001.

At start time *to* the *Measure of Belief* of the sensorial alarm of the infrared movement detector MD1 (see formula 3) is expressed by:

$$MOB \ (MD1, to):=\{ \ R \ (alarm, MD1, to), P \ (Y/A, MD1, to),$$
$$P \ (Y/A^*, MD1, to), P \ (A, MD1), score\}$$

We obtained following initial value of the measure of belief of the MD1 :
$$MOB \ (MD1, to):= \ \{ \ 1, 0.99, \underline{0.1}, 0.98, 0\}$$

Let us assume the *weight=0.001*, and that at the time *t1* this sensor provided five wrong sensorial observations of the type "FALSE ALARM". In this case its *score* is 5. Using the expression (1) we calculate the new estimate of the false alarm indicator at the time *t1*. The probability *P(Y/A*, MD1, to)* of the sensor *s* at the time *to* becomes at *t1*. i.e. after the first five "false alarm" signals:
$$P(Y/A^*, MD1, t1) = \ P(Y/A^*, MD1, to) + 5weight = 0.1 + 0.005 = \underline{0.105}$$
Using formula 1 for the sensor MD1 we calculate the alarm reliability attached to this sensor signal at evaluation time *t1*:

$$R(alarm, MD1, t1) = \frac{P \ (Y/A, MD1, t1.) \ . \ P \ (A, MD1)}{P(Y/A, MD1, t1) \cdot P(A, MD1) + P(Y/A^*, MD1, t1) \cdot P(A^*, MD1)}$$

The reliability of the alarm provided at the time *t1* by the movement detector MD1 decreases due to the false alarms signals, and becomes:
R(alarm, MD1, t1)= 0.9023

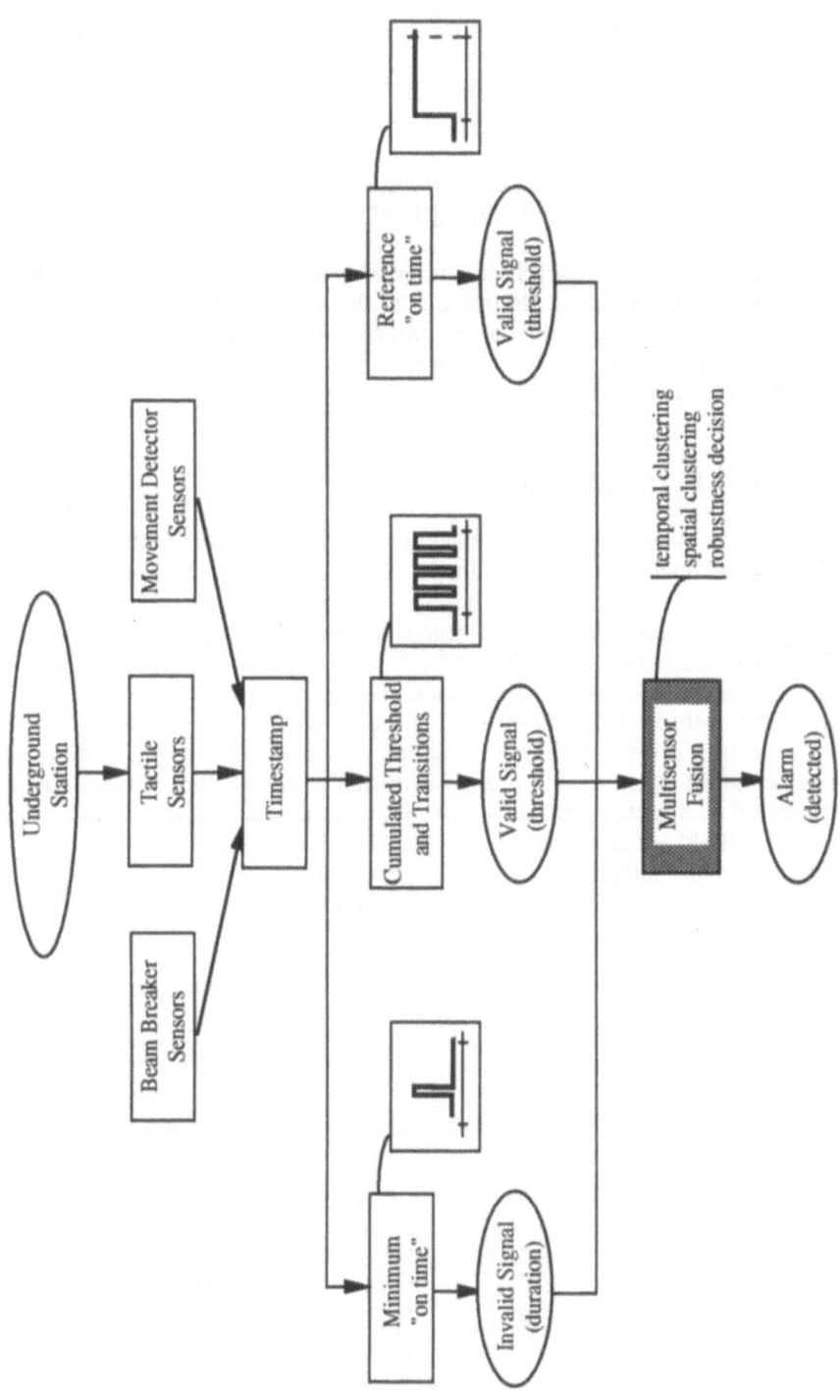

Fig. 2: Multisensor Data Fusion

The *Measure of Belief* of the infrared movement detector MD1 becomes at the time *t1*:

$$MOB \text{ (MD1, t1)} := \{ R \text{ (alarm, MD1, t1), } P \text{ (Y/A, MD1, t1),}$$
$$P \text{ (Y/A*, MD1, t1), } P \text{ (A, MD1), score(t1)} \}$$

and after using the reliability indicators values at time *t1* we obtain:

$$MOB \text{ (MD1, t1)} := \{ 0.9023, 0.99, \underline{0.105}, 0.98, 5 \}$$

This scoring scheme can be used in a similar way in order to obtain a Measure of Belief related to the "missing alarms" due to the sensor failures. The dynamic calculation of the recent reliability of the multisensorial alarm becomes in this case very complex, and can be performed in a similar way for several hierarchical levels of logical sensors.

4. Hierarchical Integration

The bottom-up fusion of the sensorial data in the DIMUS system is illustrated in figure 2. Several classes of environment observations are delivered by the redundant and diverse sensors, which have overlapped areas of observation (e.g. the tactile arrays T1 and T2, and the infrared movement detector 1 in figure 1). The observation time is added in form of a "timestamp" to the sensorial data in order to allow the ordering of the sensorial data in the temporal domain.

The invalid sensorial data is selected based on the comparison of the signal duration with an a priori specified "minimum on-time". The signal validity is determined additionally by the comparison of the signal duration with an a priori specified "reference on-time". The number of state transitions and the cumulated threshold are also used in the selection of valid signals.

The hierarchical integration of the sensorial observations is performed step by step, and several logical sensors are constructed (see figure 3).

MOVE-TOUCH 1	:=	tactile array T1 and	movement detector 1
MOVE-TOUCH 2	:=	tactile array T2 and	movement detector 1
MOVE-TOUCH 3	:=	tactile array T3 and	movement detector 2
MOVE-TOUCH 4	:=	tactile array T4 and	movement detector 2
ENTER	:=	photocell P1 and	photocell P2
GRID	:=	photocell P3 and	P4 and P5

ENTER-TUNNEL	:=	ENTER and MOVE-TOUCH 1
IN-TUNNEL	:=	MOVE-TOUCH 1 or MOVE-TOUCH 2
		or MOVE-TOUCH 3 or MOVE-TOUCH 4

TRACK	:=	(ENTER-TUNNEL or IN-TUNNEL) and then GRID
TUNNEL	:=	ENTER-TUNNEL or IN-TUNNEL
PROHIBITED-AREA	:=	TRACK or TUNNEL

Fig. 3: Hierarchical Integration

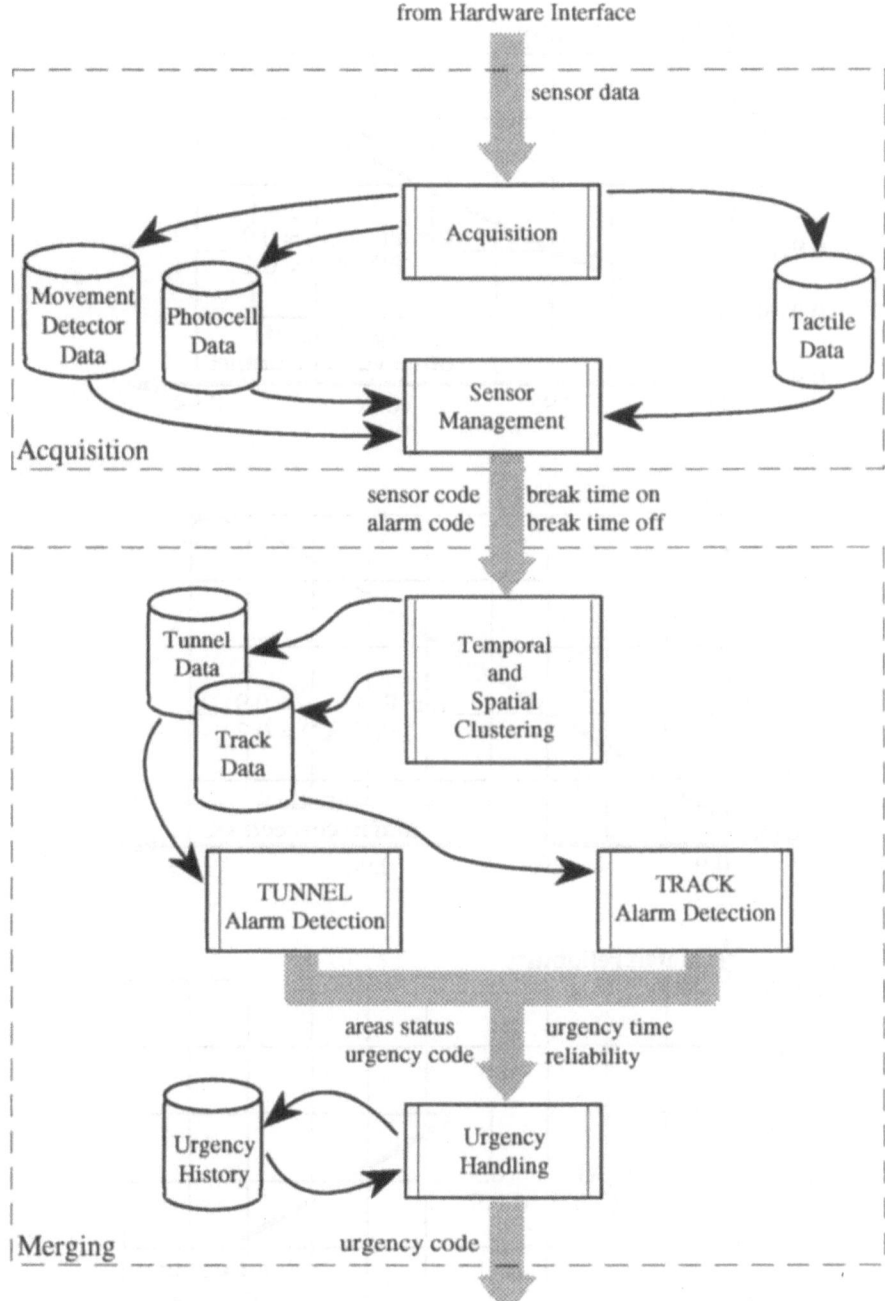

Fig. 4: Multisensor Fusion

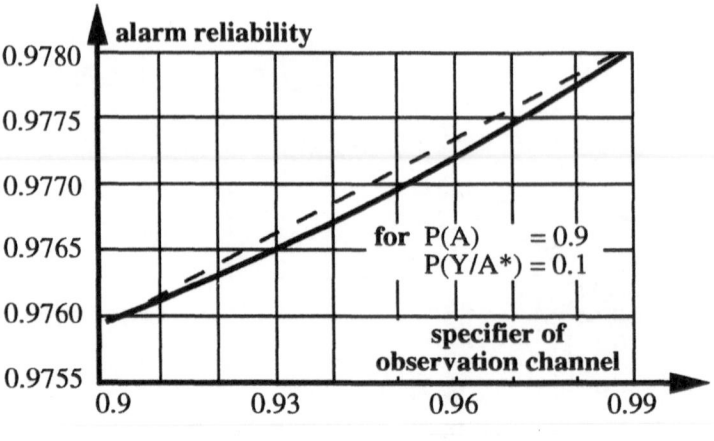

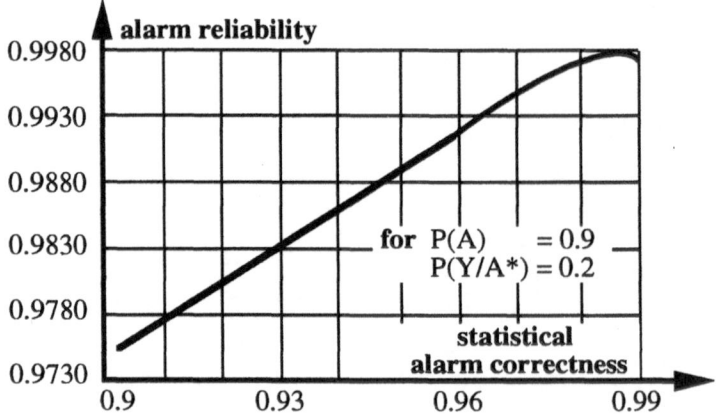

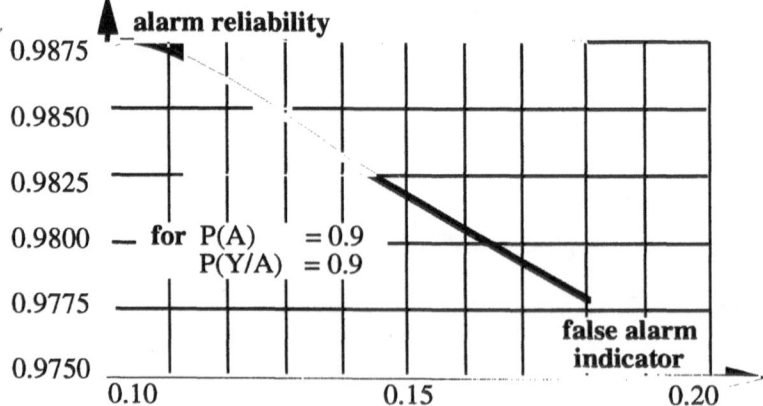

Fig. 5: Alarm reliability **R** *(alarm)* as a function of statistical alarm correctness, observation channel specifier, and false alarm indicator

The temporal and spatial clustering of the valid signals is followed by the classification of the fused signals using an urgency mechanism and a robustness decision mechanism with a weighted voting scheme.

The high-level integration of the sensorial data into the logical sensors MOVE-TOUCH 1, MOVE-TOUCH 2, MOVE-TOUCH 3, MOVE-TOUCH 4, ENTER, GRID is followed by a next integration step. In the second integration step the data obtained from the logical sensors is fused in order to construct high level logical sensors (e.g. ENTER-TUNNEL, IN-TUNNEL, TRACK, TUNNEL, PROHIBITED-AREA).

5. Conclusions

An overview of data and processes of the multisensor fusion implemented in the DIMUS system is presented in figure 4. Two main components can here be distinguished, data acquisition and the merging subsystems. Data acquisition is performed within a cycle of 100 msec. The sensorial data is stored in the related buffer. The merging subsystem constructs the high level logical sensors based on the stored sensorial data, and propagates the alarm observations. The alarm raised by the physical sensors is propagated hierarchically as a "hypothesis alarm" to the related logical sensors. The team of its sensorial observers may confirm this alarm, and in this case the alarm signal is propagated to the next higher logical sensor. Otherwise the alarm hypothesis gets the status of "false alarm", and as a result, the sensor reliability is decreased (see figure 5) in accordance with our *false alarm* handling scheme.

The four tactile arrays presented here have been constructed during the first phase of the DIMUS project, in cooperation with the company INTERLINK [10], and have been installed in an underground station in Genoa, Italy. Each tactile array consists of 25 Force and Positioning Sensing Resistors [10]. The marketing of this product is intended.

The experimental measurements of the installed multisensors confirm the robustness of this bottom-up data fusion approach.

References

1. Allen P.K., Bajcsy R., "Object Recognition using Vision and Touch", Proc. 9th Int. Joint Conf. Artificial Intelligence, pp.1131-1137, Los Angeles, USA, August 1981.
2. Allen P.K., Bajcsy R., "Two Sensors are better than one: Examples of Vision and Touch", 3rd Int. Symp. Robotics Research, MIT Press, pp.48-55, Gouvieux, France, 1986.
3. Allen P.K., "Robotic Object Recognition using Vision and Touch", Kluwer, Boston, 1987.
4. Allen P.K., "Integrating Vision and Touch for Object Recognition Tasks", Int. J. Robot. Res., Vol.7, No.6, pp.15-33, 1988.
5. Bar-Shalom Y. (editor), "Multi-Target Multi-Sensor Tracking", Artec House, 1990.

6. Didocha R.J. et al, "Integration of Tactile Sensors and Machine Vision for Control of Robotic Manipulators", Proc. 9th Conf. Robotics, pp.37-71, Detroit, USA, 1987.

7. Durrant-Whyte H.F. et al, "A modular, Transputer-based Architecture for Multi-Sensor Data Fusion", Proc. 2nd Int. Conf. Applications of Transputers, IOS Press, 1990.

8. Stansfield S.A., "Visually-aided Tactile Exploration", Proc. IEEE Int. Conf. Robotics and Automation, pp.1487-1492, Raleigh, USA, March 1987.

9. Stansfield S.A., "Primitives, Features, and exploratory Procedures building a Robot Tactile Perception System", Proc. IEEE Int. Conf. Robotics and Automation, pp.1274-1279, San Francisco, USA, April 1986.

10. Hagen J., Witte M., "Force and Position Sensing Resistors: An Emerging Technology" (published in this volume)

On Tracking Edges

Massimo Tistarelli
University of Genoa
Department of Communication, Computer and Systems Science (DIST)
Via Opera Pia 11A - 16145 Genoa, Italy

Abstract

The tracking of contours extracted from image sequences is investigated. The algorithm is based on the fusion of intensity edges and motion information (extracted from optical flow) to infer the structure of objects in space. As far as the edge tracking process is concerned it is rather general and can be applied to any kind of "ego-" or "eco-" centric motion. Furthermore, in the case of ego-motion the constraint imposed by active motion of the camera can be exploited. Within this framework in order to facilitate the measure of the navigation parameters, a constrained egomotion strategy was adopted in which the position of the fixation point is stabilized during the navigation (in an anthropomorphic fashion). This constraint reduces the dimensionality of the parameter space without increasing the complexity of the equations.

The edge tracking causes an accumulation of the errors, relative to each instantaneous displacement, up to the global cumulative image displacement. These errors can be evaluated and reduced using a simple procedure, in which the computed image displacement is combined with a prediction based on the contour trajectory extrapolated from the preceding frames.

Experimental results on real image sequences are presented.

1. Introduction

The tracking of image features in image sequences constitutes one of the more interesting topics in todays computer vision. The estimation of correspondences in images of a sequence, has much in common with the matching problem in stereo [1,2,3,4,5,6] . The basic differences between the two computational problems arise from the fact that the motion of the camera can be actively controlled in case of egomotion whereas in stereo vision it is fixed by the stereo geometry (it is only a virtual motion). For this reasons motion can be analyzed by a *tracking* strategy while stereo must be based on *matching* procedures [7,8,9,10,11,12,13,14]. The measurement of the instantaneous velocity can be used to solve ambiguities in the correspondence problem. Considering *distant* frames in time, the matching of corresponding contours can be faced by matching edge points in successive images and then tracking the points over time.

Interesting results have been presented by Kanade [15] for the case of constrained camera motion. In this work the camera was performing a fronto-parallel translation with respect to the image plane, either along the horizontal or vertical axis of the camera Cartesian coordinate system. Both camera motions produced parallel image velocity vectors with the focus of expansion at infinity. Vertical or horizontal edges (orthogonal to the direction of motion) were extracted and matched over time. In order to facilitate the matching and also reduce the probability of false matches, the grabbing frequency of the images was very high. In this way the very small differences between successive frames were obtained. An implementation of the Kalman filter was used to accumulate the velocity estimates over time. In order to make the kalman equations linear, the accumulation process was performed on the final depth estimates.

Faugeras et al. [16] and also Ayache and Faugeras [17] presented an implementation of the Kalman filter used to match edge segments and successively accumulate depth estimates. The motion of the camera was unconstrained and the images sampled at low frequency. The matching was performed imposing a rigidity constraint to three-dimensional line segments. An interesting study is performed to design an optimal representation of a straight edge segment and the associated uncertainty in space. The representation was of fundamental importance in defining the constraints for the matching process. A general version of the Kalman filter (Extended Kalman Filter) devised to handle non-linear processes, is used to build and update a 3D description of the environment.

Bandopadhay et al. and Aloimonos et al. [18,19] and also Sandini and Tistarelli [20,21] proposed a particular motion strategy in which the observer tracks the point in space which projects on the image center (the fixation point) during the movement. This motion strategy is similar to that used by the human oculo-motor system during active motion. The tracking strategy, which is possible in the case of egomotion, can be simplified by actively controlling the navigation parameters (this is equivalent to a reduction of dimensionality of the parameter space In fact from the knowledge of the egomotion parameters the two-dimensional direction of motion (i.e. the projection of the direction of motion on the image plane) can be computed uniquely. The amplitude of displacement (or the amplitude of the velocity vector) must be computed from the visual data. But there is more of it:

- the movement can be visually guided, simplifying the motion control.
- and almost the same portion of the scene can be kept within the observer field of view, without modifying the optics.

In order to increase the reliability of the measures the computation of displacement can be made only at the edge points [14]. The image velocity can be either computed from the combination of proprioceptive and image data [21] or applying a general method for the computation of the optical flow based on local constraints on the local structure of the two-dimensional motion field. The instantaneous velocity is computed at the image contours and used to establish an explicit (and discrete in space and time) correspondence between contours in successive frames. A global trajectory followed by the contour points is obtained by linking the estimates over time.

The matching of image pairs, iterated over time, causes a propagation of the errors in the estimated 2D displacements. Discretization, occlusions, image noise and poor differential methods represent an unavoidable source of errors in the processing of images from real scenes. A way to take into account these errors is to perform a regularization of the input data, for example band-limiting (either in space or/and time) the frequency of the signal (i.e. the flow of images) [22,23]. This approach can be applied, filtering the images with a band-pass filter in 2 or 3 dimensions; some examples have been already proposed [24,25,26].

Regularization techniques can be applied, beneficially, also in the estimation of visual motion, but, inevitably, fail when multiple sources of errors occur (for example related to the environment or the acquisition device). In this case it is more suited to evaluate explicitly the *amount of reliability* , or conversely the *uncertainty* embedded in the estimated variables as function of the input data and the parameters used by the algorithm. Following this approach an uncertainty value can be associated to any measurement performed, which represents the reliability of the measure. In this framework the crucial problem is how to reduce/eliminate measurement errors. Apart from improving the accuracy of the algorithm and/or of the parameters used, there are two practical solutions:
- to use multiple sensor modalities, combining the output of each of them according to the relative uncertainty [27];
- to perform repeated measurements of the same quantity, merging all the results according to their errors [28].

The two approaches are practically different but are based on the same concept. In the case of motion it is always possible to improve the accuracy in the estimation, for example, of depth or the trajectory of a point in space, processing more frames. On the other hand, at any time, it is possible to use past measurements, e.g. *the experience* , to improve the current parameter estimation [17], for example making a prediction.

This mechanism is implemented in two steps: in the first phase (*bootstrap*) the optic flow is computed for each image and, after a given time span (for example 5 frames), also depth is computed; in the second phase (*steady state*), the optic flow is computed for each frame, along with a prediction of the velocity field from trajectory extrapolation. After the bootstrap phase, for each new image, the prediction is combined with a new measurement of the flow field.

2. Computing 2D velocity and edge matching

The analysis is performed at the image contours only. They are extracted from the zero crossings (ZC) of Laplacian of Gaussian ($\nabla^2 G$) filtered images. Even though, in the past few years some alternatives have been proposed to the $\nabla^2 G$ approach, claiming that ZC are poorly localized in space, good motion and depth estimates can be still obtained. The localization accuracy in the edge detection is not mandatory, in fact, if the successive processes do not rely on accurate point localization. Moreover, better performances can be always obtained improving the edge localization (for example using a coarse to fine matching strategy to estimate the displacement of ZC at high resolution) without modifying the computational schema.

The instantaneous image velocity is computed by solving an overdetermined system of equations at the edge points only. The algorithm imposes both the constancy of the image brightness over time (constant amplitude of velocity) and of the intensity gradient (constant direction of velocity) originating three equations in the unknowns $\vec{V} = (u, v)$. They are solved, in closed form, with a standard least squares method:

$$\vec{V}(x, y, t) = (A^t A)^{-1} A^t \vec{b} \tag{1}$$

$$A = \begin{bmatrix} \dfrac{\partial I}{\partial x} & \dfrac{\partial I}{\partial y} \\[2ex] \dfrac{\partial^2 I}{\partial x^2} & \dfrac{\partial^2 I}{\partial x\,\partial y} \\[2ex] \dfrac{\partial^2 I}{\partial x\,\partial y} & \dfrac{\partial^2 I}{\partial y^2} \end{bmatrix} \qquad \vec{b} = \begin{bmatrix} -\dfrac{\partial I}{\partial t} \\[2ex] -\dfrac{\partial^2 I}{\partial x\,\partial t} \\[2ex] -\dfrac{\partial^2 I}{\partial y\,\partial t} \end{bmatrix}$$

The measurement accuracy closely depends on the properties of the 2D motion field and on the conditioning of the Hessian matrix (under this point of view this method seems more robust then the one proposed by Girosi et al. [29] because one more constraint equation is used).

The velocity estimates are effected by errors if the motion field is not locally constant, like in case of dilation,rotation or shear. For the adopted constrained egomotion the theoretical error is obtained by differentiating the velocity equations:

$$\begin{bmatrix} \Delta u \\ \Delta v \end{bmatrix} = \begin{bmatrix} \dfrac{\partial u}{\partial x}\Delta x + \dfrac{\partial u}{\partial y}\Delta y \\[2ex] \dfrac{\partial v}{\partial x}\Delta x + \dfrac{\partial v}{\partial y}\Delta y \end{bmatrix} \tag{2}$$

$$\begin{cases} \dfrac{\partial u}{\partial x} = \dfrac{[D_1 - D_2 \cos\phi \cos\theta]}{Z} + \dfrac{y\,\phi - 2\,x\,\theta}{F} < 1 + \dfrac{y\,\phi - 2\,x\,\theta}{F} \\[2ex] \dfrac{\partial u}{\partial y} = \dfrac{x\,\phi}{F} + \psi \\[2ex] \dfrac{\partial v}{\partial x} = -\dfrac{y\,\theta}{F} - \psi \\[2ex] \dfrac{\partial v}{\partial y} = \dfrac{[D_1 - D_2 \cos\phi \cos\theta]}{Z} + \dfrac{2\,y\,\phi - x\,\theta}{F} < 1 + \dfrac{2\,y\,\phi - x\,\theta}{F} \end{cases}$$

where D_1 and D_2 are the distances of the camera from the fixation point at two successive time instants, ϕ, θ and ψ are the rotational angles performed by the camera during the tracking motion, F is the focal length of the camera in pixels and Z is the depth of the considered point in space. The approximation $[D_1 - D_2 \cos\phi \cos\theta]/Z = W_Z/Z < 1$ holds for all the world points which remain within the field of view of the observer throughout the sequence. This assumption is generally satisfied for all the image points for which the optic

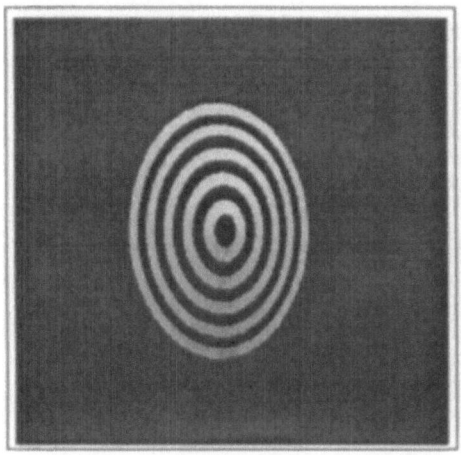 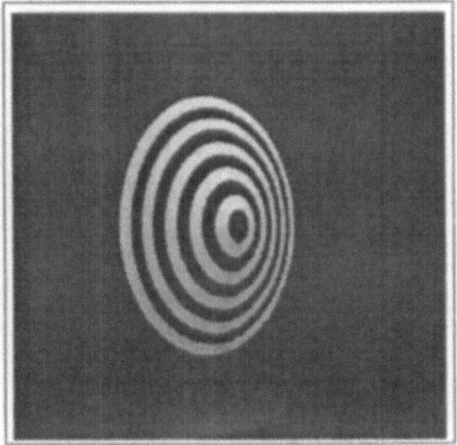

Fig. 1: First and last image of the first sequence.

flow can be computed.

The velocity estimates are used to match corresponding contour points in successive image pairs. Starting from a contour point (x_0, y_0), the matching is performed by searching for a corresponding contour in the successive image along the direction of $\vec{V}$. Even though the velocity is accurately estimated, false matches can still occurr. For this reason the corresponding contour is searched within a neighborhood of the point $(x_0 + u, y_0 + v)$ equal to the theoretical error $(\Delta u, \Delta v)$ in the velocity estimate.

A further constraint, which is used to minimize the probability of false matches, is the sampling frequency of the images. If the images are sampled at a high frequency the displacements are kept very small, reducing the searching interval.

The reliability of the matching also depends on the smoothing operated by the $\nabla^2 G$ operator. The distance between adjacent contours in the same image is not less then the parameter $w = \sqrt{2}\,\sigma$ of the Gaussian filter. Therefore, if the displacement is small and the $\nabla^2 G$ operator is sufficiently large (according to the predicted range of displacements) false matches due to spurious contours along the search direction, have low probability to occur.

A similarity measure of matched contours, based on the difference between the edge slope and orientation, allows to reject most wrong matches. In figure 1 the first image of a sequence of 15 is presented. The sequence has been acquired during a rotational motion of the camera around a cone. The fixation point was kept still during the motion. The contour map relative to the first image is shown in figure 2 (left). In figure 2 (right) the result of the edge tracking process iterated over 8 frames is shown.

The poor localization accuracy of matched contour points can be improved by adopting a coarse to fine strategy. At each resolution scale the correspondence is refined by locating the edges at higher resolution and searching for the matching using the estimate obtained at the previous resolution scale. At each step the corresponding contour, in the successive image, is searched within a window of amplitude $\frac{w}{2}$ relative to the previous estimate.

This schema allows to determine robust matches at the coarser resolution

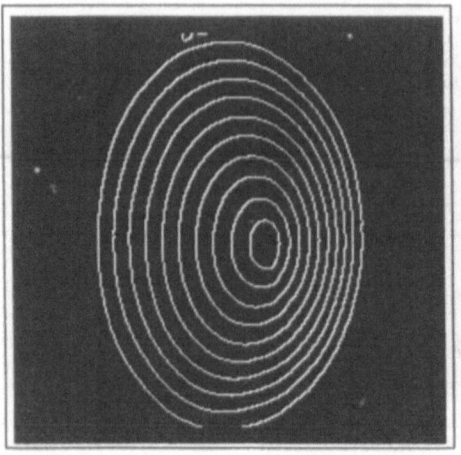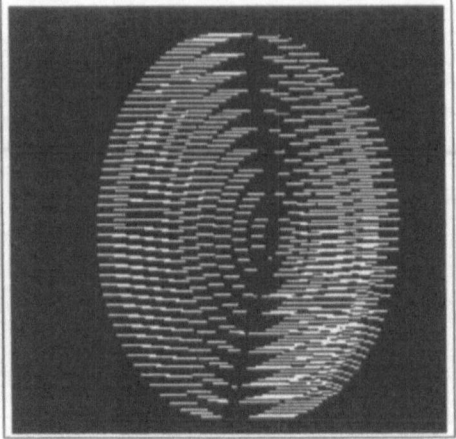

Fig. 2: Relevant contours from the first image of the sequence (left). Cumulative optical flow obtained after the edge tracking process (right).

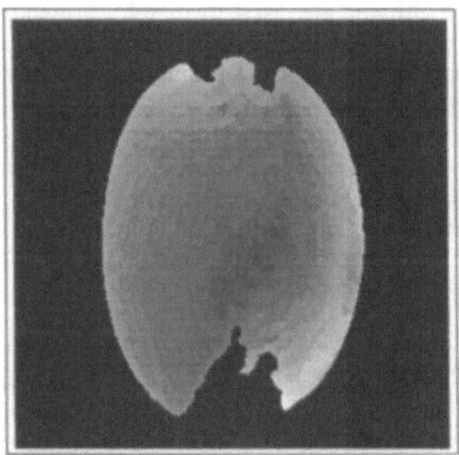

Fig. 3: Recovered range map obtained by diffusing the depth estimated at the contour points over all the image.

scale. The localization accuracy is improved by refining the matching at the successive resolution scales.

2.1 Edge tracking

After the matching process between an image pair, the edges can be tracked by iterating the process over the successive image. This schema is intrinsecally affected by measurement errors. The errors in the edge matching are cumulated over the images in the sequence. A way to bound the errors, at least within few frames, is to locally assume a linear trajectory of the edges. In this way the displacement used to perform the match is given by taking the mean direction between the displacement estimated at the previous frame and the current

189

Fig. 4: First image of the second sequence (a soda can).

velocity estimate. This assumption can be only verified locally and can not be applied to more than two frames successively. Another method is to make a prediction of the velocity at a given time, by interpolating the trajectory of the contour points in the previous frames.

A polinomial curve is used to fit the trajectory. The polinomial parameters are estimated with a least squares fitting; we choose this technique because doesn't force the polinomial to pass exactly through the trajectory points, therefore possible tracking errors can taken into account.

Among the possible interpolant curves, a second order polinomial is used, mainly for two reasons: firstly the velocity field equations are quadratic, hence a curved trajectory is expected; secondly, due to the fact that images are acquired at high sampling frequency, the trajectory is smooth and should have a limited curvature. This last assumption has been verified from experiments, made in the past on real image sequences [20,21].

From the parameters of the curve we compute, for each contour point of the current image, the coordinates of the corresponding point in the successive frame. The error of the prediction is given by the least square spread error measure.

The predicted displacement can be statistically combined with the new velocity estimate to obtain a new displacement to be used for the matching:

$$\vec{V}_* = \vec{\sigma}_*^2 \left[\frac{u_c}{\sigma_c^2} + \frac{u_p}{\sigma_p^2} , \frac{v_c}{\xi_c^2} + \frac{v_p}{\xi_p^2} \right]$$

$$\vec{V}_* = (u_* , v_*) \qquad \vec{\sigma}_*^2 = \left(\sigma_*^2 , \xi_*^2 \right)$$

$$\vec{\sigma}_*^2 = \left[\frac{\sigma_c^2 \sigma_p^2}{\sigma_c^2 + \sigma_p^2} , \frac{\xi_c^2 \xi_p^2}{\xi_c^2 + \xi_p^2} \right]$$

$\vec{\sigma}_*^2$ is the variance of the updated velocity $\vec{V}_*$, $\vec{V}_c$ is the computed velocity, while $\vec{V}_p$ is the velocity predicted from the trajectory extrapolation. The variances

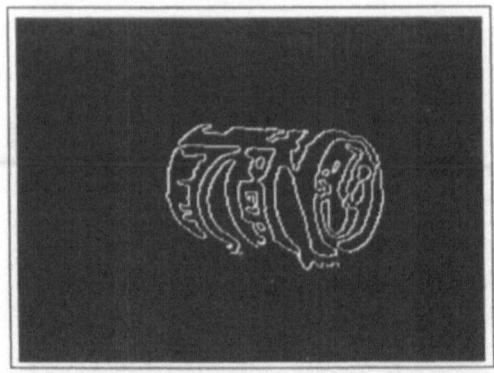

Fig. 5: Edge contours for the first image of the second sequence (above). Corrected optical flow including prediction (below).

(σ_c^2 , ξ_c^2) of the computed velocity, are estimated by differentiating the least squares solution (1), in closed form. The variances (σ_p^2 , ξ_p^2) of the predicted velocity are estimated using a standard least square χ^2 error formula.

In figure 4 the first image of a sequence of 10 of a soda can is presented. The resolution of the images is 256x256 pixels with 8 bits of resolution in intensity. The motion of the camera was a rotation around the fixation point, which was kept still during the movement (similar to the motion performed in the previous example). In figure 5 the result of the optical flow computation and the tracking process is presented. The displacement field has been obtained by integrating the computed displacement with the prediction.

3. Depth from motion

The tracking process is necessary, in the case of egomotion, to provide a larger baseline for depth estimation [31].

The distance of the objects in the scene is determined from the optic flow and the known egomotion parameters using the following equation [32]:

$$|Z| = \frac{D_f W_Z}{|\vec{V_t}|} \tag{3}$$

where $\vec{V_t}$ is the component of the image velocity vector due to camera translation: it is computed by subtracting the rotational component $\vec{V_r}$ from the whole velocity $\vec{V}$; D_f is the displacement of the considered contour point from the FOC (FOE); W_Z is the velocity of the camera along the optic axis; Z is the distance of the world point from the camera along the direction of the optic axis.

From the displacement vectors in figure 2, the depth of the edge points is computed. The recovered range map of the cone, obtained by diffusing the depth estimated at the contour points over all the image, is shown in figure 3. In figure 6 the range map for the second sequence, obtained from the optical flow in figure 5, is shown.

The distance D_f of a pixel P_i from the FOE can be expressed in the following way:

$$D_f = |FOE_x - x, FOE_y - y| = \frac{|FW_x - xW_z, FW_y - yW_z|}{|W_z|}$$

then eq. (3) can be re-written as:

$$Z = \frac{N}{|\vec{V_t}|} + W_z = \frac{M}{|\vec{V_t}|} \tag{4}$$

$$N = \sqrt{(FW_x - xW_z)^2 + (FW_y - yW_z)^2} \qquad M = N + |\vec{V_t}| W_z$$

Assumed the tracking egomotion strategy the depth function Z results :

$$Z = Z(x, y, V_x, V_y, D_1, D_2, \phi, \theta, \psi, F)$$

Considering all the state variables as Gaussian and uncorrelated, then the variance of Z is expressed, using a linear approximation, as:

$$\sigma^2_Z = N^2 \left(\frac{\partial}{\partial V_x} \frac{1}{|\vec{V_t}|} \right)^2 \sigma^2_{Vx} + N^2 \left(\frac{\partial}{\partial V_y} \frac{1}{|\vec{V_t}|} \right)^2 \sigma^2_{Vy} + \tag{5}$$

$$+ \left(\frac{\partial}{\partial x} \frac{N}{|\vec{V_t}|} \right)^2 \sigma^2_x + \left(\frac{\partial}{\partial y} \frac{N}{|\vec{V_t}|} \right)^2 \sigma^2_y + \left(\frac{\partial}{\partial F} \frac{N}{|\vec{V_t}|} \right)^2 \sigma^2_F +$$

$$+ \frac{1}{|\vec{V_t}|^2} \left(\frac{\partial}{\partial D_1} M \right)^2 \sigma^2_{D1} + \frac{1}{|\vec{V_t}|^2} \left(\frac{\partial}{\partial D_2} M \right)^2 \sigma^2_{D2} +$$

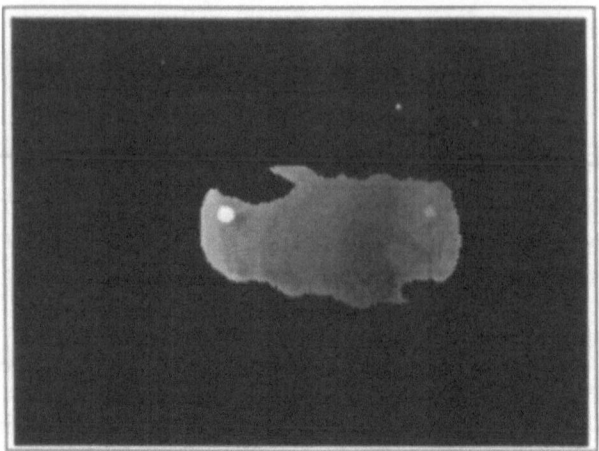

Fig. 6: Recovered range map obtained by diffusing the depth estimated at the contour points, from the optical flow in figure 5.

$$+ \left(\frac{\partial}{\partial \theta} Z \right)^2 \sigma^2_\theta + \left(\frac{\partial}{\partial \phi} Z \right)^2 \sigma^2_\phi + N^2 \left(\frac{\partial}{\partial \psi} \frac{1}{|\vec{V_t}|} \right)^2 \sigma^2_\psi$$

where σ^2_Z represents the variance of depth, σ^2_x, σ^2_y are the errors in the localization of the contour point and σ^2_F is the variance of the computed focal length of the camera expressed in pixels. σ^2_{D1}, σ^2_{D2} and σ^2_θ, σ^2_ϕ, σ^2_ψ are the variances of the known egomotion parameters of the camera (i.e. the distances of the camera from the fixation point D_1, D_2 and the rotational angles ϕ, θ, ψ). These variances depend upon the accuracy of the measurement devices, while the variance of the focal length is obtained from the characteristics of the imaging sensor (position of the image center, deviation of the optical axis etc.). The variance of the pixel position (x, y) corresponds to the error in the localization of the contour, due to the $\nabla^2 G$ filtering (which is assumed to be approximatively equal to half the standard deviation σ of the mask [33]). A better approximation can be obtained computing the statistic of the image (see for example [28]).

Bharwani et al. [34] proposed a simplified, but still general, expression of the error in depth, which only regards the errors in the velocity field, but better points out the dependence from the amplitude of velocity:

$$|\delta Z| = \frac{|D_f W_Z|}{|\vec{V_t}|^2} |\delta \vec{V_t}| \tag{6}$$

where $|\delta \vec{V_t}|$ is the magnitude of the accuracy in measuring the displacement of the image point. The accuracy in depth results directly proportional to the squared image displacement, hence, the greater the magnitude of the component $\vec{V_t}$, the greater the accuracy in depth computation.

4. Conclusion

The estimation of image velocity can be faced computing the 2D displacement of contour points. The advantage of considering the edges only stems from the great data reduction in processing the images (few hundred points versus tens of thousand in a common image). Moreover, matching edge points in successive image pairs, it is possible to estimate the cumulative displacement relative to a given number of frames. This strategy has the double advantage of allowing an accurate estimation of the displacement between image pairs, because the displacement is small, while other processes like depth computation can rely on a longer baseline. The tracking process has the disadvantage of cumulating the errors from frame to frame. A way of reducing these errors these errors is to mediate the velocity computed at each image with the predicted dislacement, obtained by extrapolating the trajectory followed by the edges at the previous frames. This process is performed by overaging each measurement with the respective uncertainty. This operation is equivalent to the application of an optimal Kalman estimator which minimizes the probability of errors. The matching between image contours is facilitated if the zero crossing are obtained filtering the images with a large $\nabla^2 G$ operator. In this case the probability of wrong matches is considerably reduced, at the price of localization accuracy. Nevertheless a good localization of corresponding contour points can be obtained by using a coarse to fine matching strategy to successively refine the estimates.

Acknowledgements

This work has been partially funded by a Basic Research Action project (P3274 - FIRST) of the European Community.

References

1. P. Morasso, G. Sandini and M. Tistarelli, "Active Vision: Integration of Fixed and Mobile Cameras", in P. Dario (ed.) *Sensors and Sensory Systems for Advanced Robots* , NATO ASI F, Vol. 43, Springer-Verlag, Heidelberg, 449-462, 1988.
2. D. H. Ballard and C. M. Brown, "Computer Vision", Prentice-Hall, New Jersey, 1982.
3. D. Marr and T. Poggio, "A Computational Theory of Human Stereo Vision", in *Proc. R. Soc. Lond.*, Vol. B 204, 301-328, 1979.
4. Whitman Richards, "Structure from stereo and motion",A.I. Memo 731, MIT A.I. Laboratory, Boston (MA), September, 1983.
5. N. J. Bridwell and T. S. Huang, "A Discrete Spatial Representation for Lateral Motion Stereo", *CVGIP* , Vol. 21, 33-57, 1983.
6. G. Sandini, P. Morasso and M. Tistarelli, "Motor and Spatial Aspects in Artificial Vision", in *Proc. of 4th Intl. Symposium of Robotics Research* , MIT Press, Santa Cruz (CA), August 9-14, 1987.
7. T. Poggio, "Visual Algorithms", A.I. Memo No. 683, MIT A.I. Laboratory, Boston (MA),1982.

8. Katsushi Ikeuchi, H. Keith Nishihara, Berthold K.P. Horn, Patrick Sobalvarro and Shigemi Nagata, "Determining grasp points using photometric stereo and the PRISM binocular stereo system", A.I. Memo 772, MIT A.I. Laboratory, Boston (MA),1984.

9. H.K. Nishihara, "PRISM: a practical real-time imaging stereo matcher", A.I. Memo 780, MIT A.I. Laboratory, Boston (MA),1984.

10. L. S. Davis, Z. Wu and H. Sun, "Contour-Based Motion Estimation", *CVGIP* , Vol. 23, 313-326, 1983.

11. C. Braccini, G. Gambardella, A. Grattarola, L. Massone, P. Morasso, G. Sandini and M. Tistarelli, "Object Reconstruction from Motion: Comparision and Integration of Different Methods", in *Proc. of II Intl. Workshop on Time-Varying Image Processing and Moving Object Recognition* , Florence (Italy), September 8-9, 1987.

12. G. Sandini and M. Tistarelli, "Recovery of Depth Information: Camera Motion as an Integration to Stereo", in *Proc. of IEEE" Workshop on Motion: Representation and Analysis* , 39-43, Kiawah Island Resort (NC), May 7-9, 1986.

13. H. H. Nagel and W. Enkelmann, "An investigation of smoothness constraints for the estimation of displacement vector fields from image sequences", *IEEE Transaction on PAMI* , Vol. PAMI-8, No. 1, 565-593, 1986.

14. E. C. Hildreth, "The Measurement of Visual Motion", MIT Press, Cambridge, (MA), 1983.

15. L. Matthies and T. Kanade, "Using Uncertainty Models in Visual Motion and Depth Estimation", in *Proc. of 4th Intl. Symposium of Robotics Research*, MIT Press, 120-138, Santa Cruz (CA), August 9-14, 1987.

16. O.D. Faugeras, F. Lustman and G. Toscani, "Motion and Structure from Motion from Point and Line Matches", in *Proc. of IEEE Intl. Conf. on Computer Vision*, 25-34, London (UK), June, 1987.

17. N. Ayache and O.D. Faugeras, "Maintaining Representations of the Environment of a Mobile Robot", *IEEE Trans. Robotics Automat.* , Vol. RA-5, No. 6, 804-819, 1989.

18. A. Bandopadhay, B. Chandra and D. H. Ballard, "Active Navigation: Tracking an Environmental Point Considered Beneficial, in *Proc. of IEEE" Workshop on Motion: Representation and Analysis"* , 23-29, Kiawah Island Resort (NC), May 7-9, 1986.

19. J.Y. Aloimonos, A. Bandopadhay and I. Weiss, "Active Vision", *International Journal of Computer Vision* , Vol. 1, No. 4, 333-356, Kluwer Academic Publishers, Boston (MA), 1988.

20. G. Sandini, V. Tagliasco and M. Tistarelli, "Analysis of Object Motion and Camera Motion in Real Scenes", in *Proc. IEEE Intl. Conference on "Robotics & Automation"*, 627-633, San Francisco (CA), April 7-10, 1986.

21. G. Sandini and M. Tistarelli, "Active Tracking Strategy for Monocular Depth Inference over Multiple Frames", *IEEE Trans. on PAMI* , Vol. PAMI-12, No. 1, 13-27, 1990.

22. T. Poggio and V. Torre, "Ill-Posed Problems and Regularization Analysis in Early Vision", A.I. Memo No. 773, MIT A.I. Laboratory, Boston (MA).

23. T. Poggio, V. Torre and C. Koch, "Computational Vision and Regularization Theory", *Nature* , Vol. 317, 314-319, 1985.
24. J. G. Bliss, "Velocity-tuned Filters for Spatio-temporal Interpolation", in *Proc. of IEEE "Workshop on Motion: Representation and Analysis"* , 61-66, Kiawah Island Resort (NC), May 7-9, 1986.
25. E. H. Andelson and J. R. Bergen, "The Extraction of Spatio-Temporal Energy in Human and Machine Vision", in *Proc. of IEEE "Workshop on Motion: Representation and Analysis"* , 151-155, Kiawah Island Resort (NC), May 7-9, 1986.
26. T. Poggio and M. Fahle, "Visual Hiperacuity: Spatiotemporal Interpolation in Human Vision", in *Proc. Royal Society London* , Vol. B 213, 451-477, 1981.
27. E. Grosso, G. Sandini and M. Tistarelli, "3D Object Reconstruction Using Stereo and Motion", *IEEE Trans. on Syst. Man and Cybern.* , Vol. SMC-19, No. 6, 1465-1476, 1989.
28. L. Matthies, R. Szeliski and T. Kanade, "Kalman Filter-Based Algorithms for Estimating Depth from Image Sequences", CMU-RI-TR-88-1, Carnegie Mellon University, Pittsburg (PA), 1988.
29. S. Uras, F. Girosi, A. Verri and V. Torre, "Computational approach to Motion perception", *Biological Cybernetics* , Vol. 60, 68-87, 1988.
30. B. Kamgar-Parsi and B. Kamgar-Parsi, "Evaluation of Quantization Error in Computer Vision", in *Proc. of DARPA Workshop on "Image Understanding"* , 1988.
31. D. Vernon and M. Tistarelli, "Using Camera Motion to Estimate Range for Robotic Parts Manipulation", *IEEE Trans. on Robotics and Autom.*, Vol. RA-6, No. 5, 509-521, 1990.
32. D. T. Lawton, "Processing Translational Motion Sequences", *CVGIP* , Vol. 22, 116-144, 1983.
33. A. Huertas and G. Medioni, "Detection of Intensity Changes with Subpixel Accuracy using Laplacian-Gaussian Masks", *IEEE Transaction on PAMI* , Vol. PAMI-8, No. 5, 651-664, 1986.
34. S. Bharwani, E. Riseman and A. Hanson, "Refinement of Environmental Depth Maps Over Multiple Frames.", in *Proc. of IEEE "Workshop on Motion: Representation and Analysis"* , 73-80, Kiawah Island Resort (NC), May 7-9, 1986.

Force and Position Sensing Resistors : An Emerging Technology

Jannik Hagen, Michel Witte
Interlink Electronics Europe
L-6401 Echternach, Luxembourg

Abstract

Force Sensing Resistor™ devices (FSR™) superficially resemble a membrane switch, but unlike the conventional switch, change resistance inversely with applied force. For example, with a typical FSR sensor, a human finger applying a force from 0,1 N to 10 N will cause the sensor to change resistance continuously from 400 kΩ to 40kΩ. These sensors are ideal for touch control, and may be applied where a semi-quantitative sensor is called for that is relatively inexpensive, thin (>0,15 mm), durable (10.000.000 actuations), and environmentally resistant. These sensors can be made into arrays or single elements up to 60 cm x 80 cm, and cover forces in the tens of grams to tens of kilograms range.

Force and Position Sensing Resistor™ devices (FPSR™) can sense the position and normal force of a single actuator, such as a finger or a stylus, along either a straight line (a Linear Potentiometer) or on a planar surface (an XYZ Pad). Depending on the mechanical arrangement, positional resolution of 0,05 mm is possible.

1. Introduction

Force and position sensing are integral to a wide range of dynamical measurements. These range from podiatric gait analysis to electronic music to computer input devices. New sensor options for the designer are the Force Sensing Resistor (FSR) and the Force and Position Sensing Resistor (FPSR).

We will first deal with a simple FSR. The construction of a typical FSR is shown in figure 1. A conducting pattern is deposited on one of the two polymer films in the form of a set of interdigiting electrodes. The electrode pattern is typically on the order of 0,4 mm finger width and spacing.

Next, a proprietary semiconductive polymer is deposited on the other sheet. The sheets are faced together so that the conducting fingers are shunted by the conducting polymer. When no force is applied to the sandwich the resistance between the interdigiting electrodes is quite high, usually 1MΩ or more. With increasing force, the resistance drops, following an approximate power law. A typical plot of resistance versus force is shown in figure 2.

Note that, unlike a conventional load cell or strain gauge, the FSR resistance changes by nearly 3 decades.

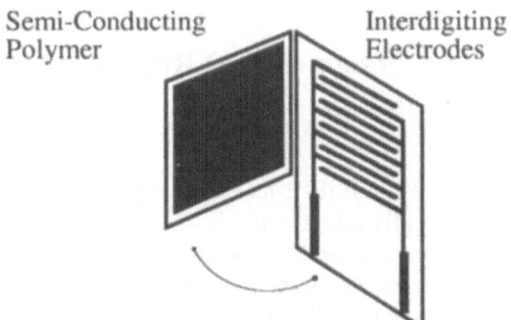

Fig. 1: Structure of a FSR™

With proper mechanical arrangement, repeatability of this curve cycle-to cycle is better than ± 2 % over a specified force range. For the device from which the data in figure 2 was obtained (a 2,5 cm diameter circular FSR), the specified force range was 200 g-10 kg. Device-to-device variation is typically ± 15 % at 10 kg.

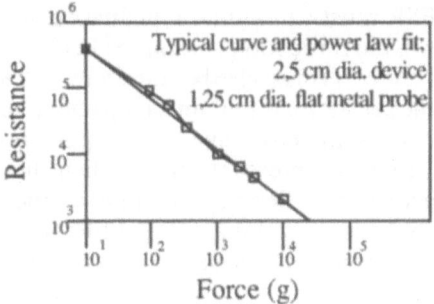

Fig. 2: FSR™ Force/Resistance Characteristic

The curve of figure 2 does not show forces above a 10 kg load. At higher forces, the force/resistance does not vary strongly with force. The saturation force is a function of the ratio of the area of the applied force to the spacing between the FSR conductive interdigiting electrodes. As we will discuss, the finer the lines and spaces, for a given area of applied force, the higher the saturation force. This saturation force can be tailored in the range between 3 and 50 kg.

FSRs can be fabricated in various sizes, from 0,5 to 4800 cm^2, as single sensors or as arrays. The resistance range can also be tailored to specific applications. Varying the force range is also possible, but is best accomplished in the mechanical design. Zero travel is also a value added feature of the FSR. However, where tactile feedback is desired, elastomeric overlays or molded domes can be used to provide some travel or a tactile "snap".

The thickness of an FSR is governed by design requirements and constraints such as desired sensitivity, presence of overlays, and specified flexibility. Nearly all FSR designs to date have been in the thickness range of 0,1-1 mm.

Compared to piezoelectric transducers, the FSR is a slow device (typical mechanical rise time of 1-2 ms), and is relatively insensitive to vibration and acoustic noise pickup.

2. Effect of Mechanical Design on FSR Response

a) Area Effects

The force/resistance response of an FSR is an extremely sensitive function of the manner in which it is mechanically addressed.

The FSR is not a true force sensor that gives a constant reading at a constant force, independent of the area over which the force is applied, or its distribution. It is not a true pressure sensor that gives, with the same constant force, a reading which is inversely proportional to the area of the applied force.

In actuality, the FSR lies somewhere between a force and a pressure transducer. A typical FSR will show a resistance that varies roughly as the reciprocal of the sqaure root of the area of the applied force. This holds true under the condition where the force footprint is smaller than the FSR active area, and large compared to the spacing between the conducting fingers.

The sensitivity of the FSR resistance to the area and distribution of the force means that either the FSR must be used as a qualitative sensor, or that by proper mechanical arrangement, the force footprint can be held constant in area, position and distribution. Other tradeoffs must also be considered in the actual sensor design; for example, tailoring the sensor for minimum creep under load conflicts with some application requirements that the FSR have a very large no-load resistance.

The FSR can be used as a pressure sensor when the applied force is large compared to the FSR active area. Semi-quantitative biomedical gauging has been accomplished by orthopedists attaching small FSRs to various body parts in configurations such that the force is constant across the sensor active area.

b) Actuator Characteristics

The compliance of the force actuator, which charges the parts of the fingers that actually contact and transfer force to the FSR, is also a key issue. Frequently, a rubber or other elastomeric overlay is placed over the part to help spread the force out, extending the dynamic range.

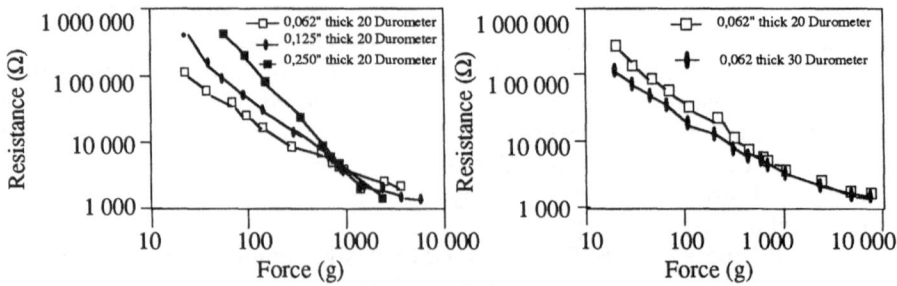

Fig. 3: Effect of Overlay Thickness & Hardness

Figure 3 shows how a typical force/resistance characteristic is changed by the use of overlays of varying thickness and hardness (or durometer, Shore A). Note that the greatest effect is seen at low to intermediate forces.

c) Conductor Design

Another key element in proper FSR sensor design is the fineness of pitch of the conductive fingers. For a given area, the finer the pitch (or "space and trace"), the greater the number of fingers actuated. The effect of the greater number of shunted fingers can be seen to increase the dynamic range of the device. In other words, the power-law characteristic is maintained over a greater force range (i.e., linearity on a log-log plot).

The trade-off here is cost. With a finer space and trace, quality assurance inspection takes longer and the rejection rate is higher. This needs to be balanced against the real-world requirements of a given design.

3. Device Durability

The FSR is a rugged, durable device. The temperature range of our standard devices extends to 85° C, continuous. Figure 4 shows the results of repeated use. For these data, a 2,5 diameter circular FSR was placed in a cycling force tester. A 55 N force was applied over ca. 1,5 cm^2, through a 3 mm thick 45 Shore A rubber foot. The force was applied and released at a 2,5 Hz rate, with a 50 % duty cycle. A small change toward lower resistance is observed after 10.000.000 cycles; however, this represents less than 5 % deviation (logarithmic) from the new part characteristic.

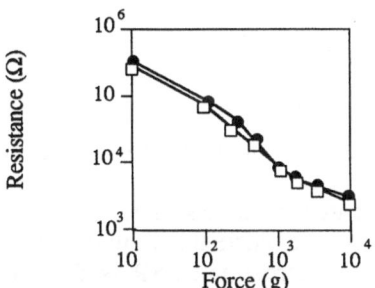

Test conditions : 0,5" dia. circular probe
10 million strikes at 52 psi; 0,5 sec/repetition

Fig. 4: FSR™ Longevity Test

4. Electrical Interfacing

As we have seen, the FSR changes resistance dramatically with applied pressure. Additionally, its impedance is nearly purely resistive. These properties make FSR electrical interfaces extremely simple. Unlike strain-gauge sensors with their low $\Delta R/R$, no bridge is needed in FSR circuits, and the signals are usually in the 0-5 volt range.

Two general rules must be kept in mind, however: first, the FSR force-resistance response characteristic is a power law, so it may make sense to measure the logarithm of resistance changes; second, the maximum permissible device current is about 1 milliamp per cm² of applied force. Typical FSR current excitation is in the tens of microamps. You can use the FSR to control larger loads by using suitable buffer circuits. The most unpredictable part of the FSR/force resistance characteristic is the pressure range under 100 g/cm². If it is necessary to measure small forces in that range, you can preload the FSR with 100-200 g/cm², and measure the change in resistance when the small load is applied. At a somewhat higher part cost, high sensitivity can be designed in, but it is generally more economical to achieve this in the mechanical interface.

The dynamic range of the FSR simplifies electrical interfacing. For instance, a simple force to frequency converter is shown in figure 5. In this circuit, the FSR is used as a feedback element around an inverter with the time constant set by the FSR resistance and the capacitor. At zero force, the FSR resistance is very high, and the oscillator does not run. With increasing force, the output repetition rate is linear through the sensor. A great deal of control of the force/frequency curve is possible by including other elements in the feedback system. For example, bypassing R1 with a capacitor causes the curve to be steeper at higher forces; connecting a large value resistor in parallel with the capacitor C quenches any tendency to oscillate at low applied forces.

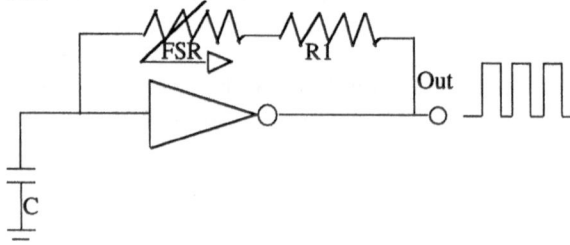

Fig. 5: A Simple Force-to-Frequency Interface

Analog interfaces are also quite simple. For example, the FSR can be placed in series with a current source (current kept within the maximum FSR rating). The voltage divider measured across the FSR is then related to the applied force. Alternately, the FSR can be used as one element in a voltage divider, with a fixed resistor as the other element. A voltage is applied to the divider, and the output voltage, taken from the resistor/FSR junction, is measured (Figure 6.).

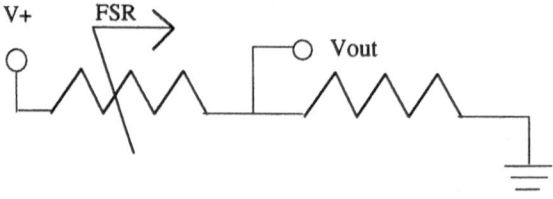

Fig. 6: Analog Interface

This type of interface is quite adequate for qualitative force sensing (for example, a touch panel). Precision measurements, however, are difficult, due to the shape of the power law curve. For higher precision measurements, it is usually ost economic to go to the digital domain as soon as possible so that the log/log characteristic of the device can be translated to something more linear. If a design calls for a measurement of an impact (for example, a data entry keypad adhered behind a rigid plate) the FSR can be placed in a voltage divider capacitively coupled to the succeeding stages. This eliminates any offset problems due to a preload. In the application just cited, denting the keypad protective plate with a hammer did not affect the operation of the pad; the offset created by the constant resistance of the FSR under the dent was blocked by the coupling capacitor.

Force and Position Sensing Resistors

Two basic types of FPSRs are available, the Linear Potentiometer (FSR-LP) and the XYZ Pad. The FSR-LP, besides being force sensitive, measures the position of an applied force along its sensing strip. The XYZ Pad measures position in a plane.

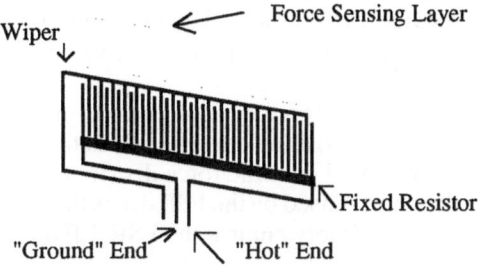

Fig. 7: Structure of a FSR-LP

Generally, a voltage is applied between the Hot and Ground ends of the fixed resistor strip. When force is applied to the Force Sensing layer, the wiper contacts are shunted through that layer to one of the conducting fingers of the resistor strip. The voltage read from the wiper is thus proportional to the distance along the strip that the force is applied. An equivalent circuit for this arrangement is shown in Figure 8. The wiper series resistance varies with force.

To sense force, a resistance measurement is made between the wiper terminal and either the Hot or Ground end (or both connected together) of the fixed resistor strip.

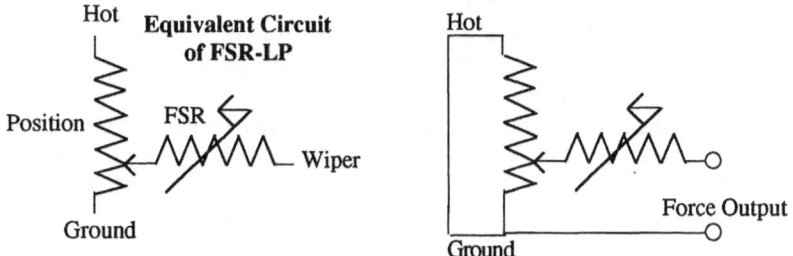

Fig. 8: Equivalent Circuit of FSR-LP

It is obvious that force and position measurements are not totally independent. However, the position measurement can be made unambiguously if the
measuring device draws negligable current ($< 1\mu A$) so that there is no voltage drop
across the wiper resistance. For example, an LF411 or similar low V_{os} and i_{bias} device
is suitable for this application. Also, the contact between the wiper and the resistor
is momentary - some sort of sample and hold must be used in the inteface. Connection
of a small capacitor between wiper and ground is usually sufficient.

Force measurements are not quite so straight forward, but good approximate
measurements can be obtained by shorting the two fixed resistor ends (Hot and
Ground) together, and measuring the resistance between the combined leads and the
wiper. Some error can result from the fixed resistor being in series with the Force
Sensing Resistor unless it is compensated for in the product design.

For example, if the force is applied at the middle of the FSR-LP, the FSR part
will have an additional resistance of 1/2 of the total fixed resistance, because the
resulting middle contact effectively parallels the two halves of the fixed resistor.
If, on the other hand, force is applied to either end of the device active area, the fixed
resistance is essentially shorted out. This problem can be mostly overcome by
making the FSR resistance high compared to the potentiometer resistance, making
low-current position measurement a must. The force measurement error can also be
subtracted out with a simple analog circuit or, given a position measurement,
compensated for in software.

With the above topologies, force measurement is of one point only. It can be
shown that the measured force position corresponds to the barycentric positon, that
is, a positional average weighted over the force distribution. For the common case
of a finger actuation, the point sensed by the FSR-LP is the center of the area covered
by the fingertip; the user can finely control the FSR-LP output by a gentle rock of
the finger.

Positional Resolution and Accuracy

Resolution of a measurement device must be distinguished from accuracy. Resolution refers to the smallest change in position that can be detected. The positional
resolution of a FSR-LP depends on the width of the applied force distribution and
the fineness of the conductive fingers on the FSR-LP. It is typically in the 10-100
micron range.

Resolution can be approximated by $\Delta x = 2w_s^2/w_f$, where w_s is the width of the
conductive fingers and w_f is the width of the applied force. This approximation
assumes a relatively constant force across the force footprint. A typical number is:
$\Delta x = (2)(0,5 \text{ mm})^2/(15 \text{ mm}) = 0,033 \text{ mm}$.

In practice, it is easy to achieve 50 counts per cm with a finger as the actuator.
The resolution could be increased by decreasing the width of the conductive fingers
and spaces; the trade-off is a more expensive part.

Positional accuracy refers to the absolute knowledge of a point's position, as
opposed to resolution, which refers to the relative knowledge of position. The
absolute positional accuracy of an FSR-LP is about $\pm 3 \%$ to $\pm 5 \%$.

The XYZ Pad

As we have seen, the FSR-LP gives measurements of normal force and position along a line. It is often desirable to measure the position of an applied force in a plane (e.g., a graphics input pad for a computer).

If linear position of a point is measured in two orthogonal directions, then the position of the point on a plane is completely specified. Conceptually, by placing two FSR-LPs back-to-back and perpendicular to each other, one can measure the position of a force on a plane, as well as the magnitude of the force. The XYZ pad is so called since it can measure plane coordinates (X and Y) and normal force (Z).

Double-sided FSR Layer

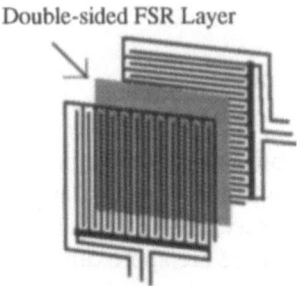

Fig. 9: Structure of an XYZ Pad

Figure 9 shows the construction of an XYZ pad. Note that an unambiguous position can only be measured for a single applied force; multiple contacts will have degeneracies (that is, a non-unique set of solutions) in force-position measurements. This is not a problem for, e.g., graphic pads, but it does mean that the XYZ Pad cannot be used for complex pattern recognition.

Fusing Two Views using Object Motion

David Hogg and Adam Baumberg
School of Computer Studies
University of Leeds, United Kingdom

Abstract

A method is described for establishing plausible point correspondences between two widely separated views using statistical regularities in the motion of objects discovered through extended observation. Strong assumptions about the motion activity in a scene are required. These are approximately satisfied in many kinds of pedestrian scene.

1. Introduction

The correspondence problem for binocular stereo or structure from motion is a well researched problem in machine vision for which there are several effective solutions (e.g. Pollard, Mayhew and Frisby, 1985; Nishihara, 1984). Most approaches presuppose views are similar to one another, ensuring local pictorial structure is comparable. In contrast, relatively little work has considered the related correspondence problem dealing with views that are widely separated. This paper describes a method for establishing a small set of possible point correspondences between such disparate views under the following special circumstances:

- Long image sequences are available from each view depicting large numbers of objects moving along discernible pathways.
- The motion of these objects is planar (i.e. objects move on a *ground plane*).
- All objects moving along similar pathways travel at approximately the same constant speed.

   Pedestrian scenes present a common situation in which these circumstances may arise and for which surveillance cameras often provide widely separated overlapping views.

   The correspondences obtained between views are shown to be sufficient to determine pointwise mappings between the projections in each view of points on the ground plane. Of course, only a correct correspondence will deliver the correct mapping. To see why such a mapping might be useful, consider tracking an object through a scene viewed by two cameras with overlapping fields of view (as in figure 1). We assume there is no problem detecting and tracking the object in either view independently. The difficulty comes in establishing that the objects detected by each camera are infact one and the same. This is easily achieved by noting the coincident positions of the object viewed from the first camera and the object viewed from the second when mapped into the image space of the first. This approach generalises to more complicated situations involving many surveillance cameras covering a large complex of buildings. Mappings between views covering overlapping parts of the site may be used to track objects as they pass from camera to camera.

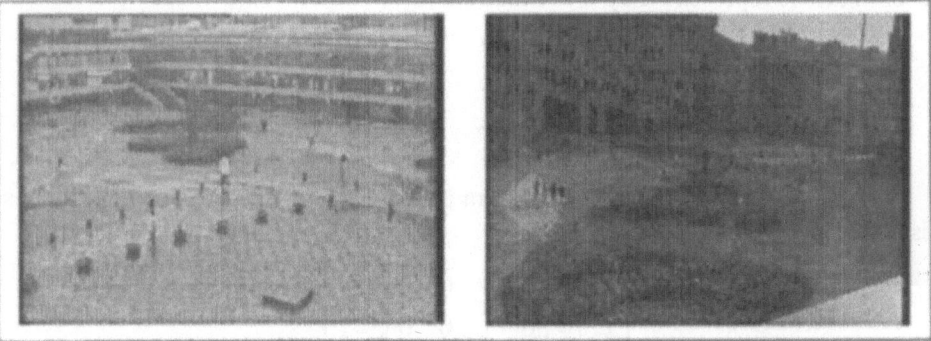

Figure 1: Two views of a pedestrian square from opposite sides

The method for establishing correspondences between views described in this paper has two useful properties:

- Motion sequences need not be obtained from each camera at the same time. This is a unusual property since it removes the possibility of exploiting simultaneous views of the same object. Rather we exploit statistical regularities in the motion of objects to establish a mapping between views.
- No knowledge is required of intrinsic or extrinsic camera parameters (e.g. the focal length, centre of projection and aspect ratio of either camera, and the relative position of cameras). We assume non-linear camera properties, such as spherical aberration, are removed by pre-processing of images.

In outline, the method works as follows. Long image sequences from both cameras are processed to find the image-plane trajectories of the projections of large numbers of moving objects. From these trajectories, pathways are detected in both image planes related to common pathways on the ground plane. For each straight section of pathway, a motion model is estimated enabling extrapolation of the motion along the line through the straight section. Combinations of two line correspondences between views with associated motion models are used to derive possible sets of four point correspondences. Finally, a mapping between camera views for the projections of points on the ground-plane is derived from each set of point correspondences.

2. Mapping between views

This section reviews the geometry of image formation to explain the form of the target mapping between views.

Consider one camera view of a ground plane with an embedded coordinate frame such that the plane is given by $Z^W = 0$. Using homogeneous coordinates for points, the rigid transformation from ground-plane coordinates to camera coordinates is represented by a 4x4 matrix

$$\begin{bmatrix} X^C \\ Y^C \\ Z^C \\ 1 \end{bmatrix} = \begin{bmatrix} r_{11} & r_{12} & r_{13} & t_x \\ r_{21} & r_{22} & r_{23} & t_y \\ r_{31} & r_{32} & r_{33} & t_z \\ 0 & 0 & 0 & 1 \end{bmatrix} \begin{bmatrix} X^W \\ Y^W \\ Z^W \\ 1 \end{bmatrix}$$

(2.1)

Perspective projection from camera coordinates to image-plane coordinates is represented by a 3x4 matrix

$$\begin{bmatrix} x_1' \\ x_2' \\ x_3' \end{bmatrix} = \begin{bmatrix} 1 & 0 & 0 & 0 \\ 0 & 1 & 0 & 0 \\ 0 & 0 & 1 & 0 \end{bmatrix} \begin{bmatrix} X^C \\ Y^C \\ Z^C \\ 1 \end{bmatrix}$$

(2.2)

The familiar projection equations are obtained by converting from homogeneous image plane coordinates (x_1', x_2', x_3') to standard Euclidean coordinates (x', y')

$$x' = x_1'/x_2' = X^C/Z^C$$
$$y' = x_2'/x_3' = Y^C/Z^C$$

Finally, an affine transformation from image-plane coordinates to pixel coordinates accommodating focal expansion, offset to the centre of projection, and non-square pixels, is represented by a 3x3 matrix

$$\begin{bmatrix} x_1^p \\ x_2^p \\ x_3^p \end{bmatrix} = \begin{bmatrix} a_{11} & a_{12} & a_{13} \\ a_{21} & a_{22} & a_{23} \\ 0 & 0 & 1 \end{bmatrix} \begin{bmatrix} x_1' \\ x_2' \\ x_3' \end{bmatrix}$$

(2.3)

A single 3x4 transformation matrix mapping directly from ground-plane coordinates to pixel coordinates is obtained by composing the three matrices (2.1), (2.2) and (2.3). Since for points on the ground-plane Z^W is always zero the third column of this matrix may be omitted giving a 3x3 matrix mapping only points on the ground plane

$$\begin{bmatrix} x_1^P \\ x_2^P \\ x_3^P \end{bmatrix} = \begin{bmatrix} t_{11} & t_{12} & t_{13} \\ t_{21} & t_{22} & t_{23} \\ t_{31} & t_{32} & 1 \end{bmatrix} \begin{bmatrix} X^W \\ Y^W \\ 1 \end{bmatrix}$$

(2.4)

This transformation is known as a *plane-to-plane projectivity* (Mundy and Zisserman, 1992). Let the plane-to-plane projectivities for the two views be T_1 and T_2 respectively. These two transformations together define a third plane-to-plane projectivity mapping projections of points on the ground plane directly from pixel coordinates in one view to pixel coordinates in the other. Thus, the mapping from the first view to the second is given by the 3x3 matrix $T_2T_1^{-1}$. Any plane-to-plane projectivity is completely specified by four point correspondences between views (no three points collinear) (Mundy and Zisserman, 1992), since there are 8 unknowns in the projectivity matrix and each corresponding pair of points gives two equations by substituting into equation (2.4). Our aim has been to devise a method for establishing the necessary four point correspondences in order to determine the projectivity between views.

3. Object detection and tracking

The inputs for the method are two TV image sequences obtained from a pair of cameras with overlapping views of a ground plane on which objects are moving. The sequences may be obtained at different times but should be long enough for statistical regularities in the motion of objects to be discernible. In our experiments we use sequences of around 5000 frames corresponding to approximately 30 minutes of live action (@ 3 frames/second).

Until the final stages in which point correspondences are sought between views, the input sequences are processed independently of one another. Several alternative methods could be used for detecting and tracking moving objects. We have chosen a simple differencing strategy in which consecutive frames of a sequence are subtracted and any 'change-points' clustered into blobs which are assumed to correspond to objects moving in the scene. By establishing correspondences between blobs detected in successive images on the basis of consistency in position and size, objects are tracked from frame to frame (for details see Hogg, 1983).

The centre point at the base of each blob (figure 2) is taken to be the projection of the instantaneous point of contact of an object with the ground plane. Successive points form trajectories describing the projection of the object's progression through the scene. To reduce the consequences of errors in the detection and tracking of objects, trajectories are smoothed and also deleted if they contain sudden changes in direction indicative of correspondence errors between blobs in successive frames.

Figure 2: Illustrating the construction of trajectories from detected blobs

Many trajectories are required to be generated from each image sequence. Figure 3 shows over 1000 trajectories detected from the two views shown in figure 1 of a moderately quiet pedestrian scene. The observation lasted about 30 minutes and was performed at different times for each view.

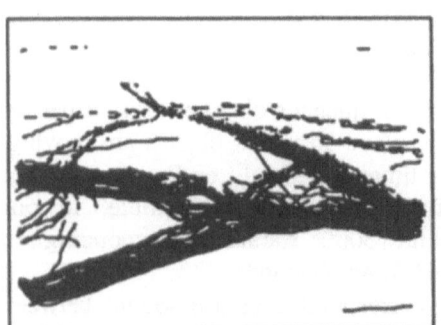 

Figure 3: Trajectories generated from two views.

4. Pathway detection

From the set of trajectories derived from each view, characteristic trajectories or *pathways* are sought along which most objects appear to be moving in the image plane. Conceptually, pathways are prototypes for clusters of similar trajectories and are represented in the same way: as sequences of instantaneous point positions. Pathways may partially overlap with one another when distinct clusters of trajectories run side-by-side for part of their length.

We have tried three different strategies for the detection of pathways, including the use of a Hough transform (see Li-qun, Young and Hogg, 1992). Only one strategy is described here. In this scheme, pathways are selected directly from the set of available trajectories. The scheme is iterative with the trajectory commanding the largest number of similar trajectories being promoted to a pathway at each iteration. Only the longest of the trajectories are assessed as candidate pathways. All trajectories similar to the promoted trajectory are themselves removed from further consideration. The details are as follows:

Assign list of trajectories to TRJS

Sort TRJS by trajectory length (i.e. by number of constituent points)

LOOP:Assign to LTRJS the first (i.e. longest) n trajectories of TRJS

For each trajectory in LTRJS, find all similar trajectories in TRJS (i.e. those which <u>on average</u> deviate from it by no more than ε pixels)

Select as a pathway the trajectory with the largest number of similar trajectories. Pathways are represented as lists of points in the same way as trajectories

Delete all trajectories similar to the selected trajectory from TRJS (including the selected trajectory itself)

If trajectories remain in TRJS then goto LOOP else stop

In our implementation, n=5 and ε=5 with a pixel image of size 180x144. Figure 4 shows the pathways chosen from the trajectories depicted in figure 3. Notice that each cluster of trajectories is represented by a pathway. Some of these clusters are relatively sparse.

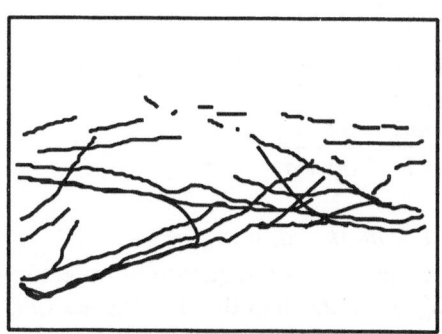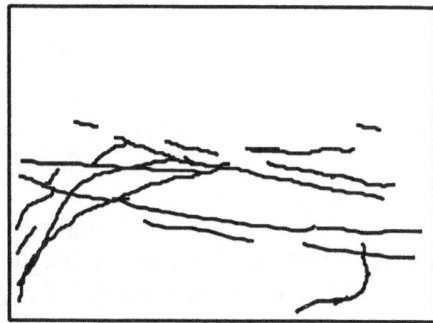

Figure 4: Pathways obtained from the trajectories shown in Figure 3

5. Modelling motion in the image

Our eventual aim is to use the pathways detected in each image plane to establish point correspondences between views and thereby compute a view-to-view transformation. To this end, a metrical model is constructed for the motion of objects along straight sections of pathway in the image planes. This will enable extrapolation of the motion beyond the ends of pathway sections and ultimately provide a mechanism for identifying identical points on the ground plane in both views.

To obtain this model we make the strong assumption that objects moving on the same pathway do so with the same constant speed across the ground plane, at least for the duration of the two image sequences. This assumption is almost certainly stronger than necessary and should be relaxed to an assumption about the statistical properties of clusters of trajectories associated with a pathway, for example, that individual objects move with constant speed (in the ground plane) from a normal distribution. We have not yet done this.

Assuming then that all objects associated with a pathway move with constant speed in the ground plane, the distance travelled by an object may be modelled as

$$s(t) = \alpha t + \beta$$

To relate this model to the projected motion of objects in the image plane, all pathways are broken into straight segments using a 2D curve splitting procedure (Duda and Hart, 1973). Under perspective projection, the 1-D coordinate s' of a point on the line through a straight section of pathway (i.e. the distance of the point along the line from some fixed point) is related by a line-to-line projectivity to the 1-D line coordinate s of the back-projected point on the corresponding line in the ground plane

$$\begin{bmatrix} \lambda s' \\ \lambda \end{bmatrix} = \begin{bmatrix} m_{11} & m_{12} \\ m_{21} & 1 \end{bmatrix} \begin{bmatrix} s \\ 1 \end{bmatrix}$$

Thus, by substituting for s and eliminating λ between the two equations, the motion of an object in the image plane is given by

$$s'(t) = \frac{m_{11}(\alpha t + \beta) + m_{12}}{m_{21}(\alpha t + \beta) + 1} = \frac{a_1 t + a_2}{a_3 t + a_4}$$

where

$$a_1 = m_{11}\alpha \quad a_2 = m_{11}\beta + m_{12} \quad a_3 = m_{21}\alpha \quad a_4 = m_{21}\beta + 1$$

A straight section of pathway of length k is composed of a sequence of trajectory points with 1-D coordinates $(s_0', s_1', s_2', \cdots, s_{k-1}')$ on the line through the section. Since TV frames are sampled at equal intervals, these trajectory points are equally spaced in time and so the subscripts may be taken as time stamps from the start of the straight section. Since only ratios between the unknown parameters a_1, a_2, a_3, a_4 are needed, let $a_1 = 1$. Each line coordinate s_{t_i}' gives one linear equation for the remaining unknowns.

$$a_2 - s'_{t_i} t_i a_3 - s'_{t_i} a_4 + t_i = 0 \qquad (5.1)$$

In general, many more than the three equations required for a solution will be available and a least squares solution is found, giving a better estimate.

This model of the projected motion is used directly to extrapolate motion along the line, beyond the ends of pathway segments, thereby predicting how an object would move in either image were it to break free from the pathway and continue moving in a straight line.

6. Finding four point correspondences

So far, straight pathway segments have been recovered from both images and a model estimated for the motion along each segment. However, no correspondence has been established between views.

In general, four point correspondences are required to estimate the required plane-to-plane projectivity between views (see Section 2). By the duality of points and lines in projective geometry, four line correspondences will do the same job. Because of the wide disparity between views, the direct use of feature points on pathways to form correspondences is unreliable. For example, the endpoints of a pair of corresponding pathways may relate to different points on the ground plane as a result of differences in the visibility from either view.

Instead we obtain four point correspondences from two line correspondences augmented by the associated motion models described above. The idea is to use these models to predict the points reached after two fixed time intervals, starting out at the intersection of the lines (pairs of lines are chosen that are not parallel). The absolute time at which motion passes through the intersection point on each line is obtained from equation (5.1). Since the underlying motion is along the ground plane, the extrapolated points computed in each image should correspond to the same point on the ground (figure 5). This procedure gives four point correspondences between images from which the required projectivity is generated.

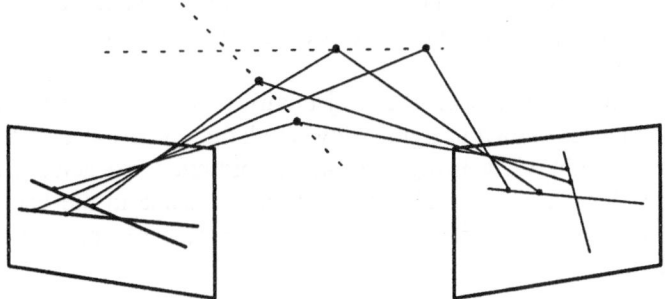

Figure 5: The construction of four point correspondences.

Figure 6 shows two correct line correspondences between the two views for which pathways are shown in figure 4. The two constructed points on each line are shown as dots. One of these points for the left view is outside the image boundary and is not shown.

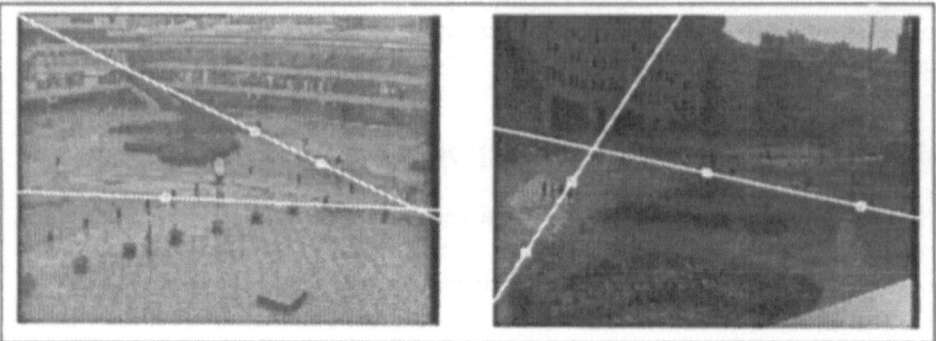

Figure 6: Two line correspondences and constructed points.

For each constructed point, two criteria are used to ensure the direction of motion from the intersection points is consistent between images. Firstly, the direction should be towards the straight section of pathway giving rise to the line in the first place. If this direction is undefined because the section of pathway straddles the intersection, the direction chosen is towards the centre of the image. Both criteria (especially the second) may lead to incorrect results and a safer approach might be to consider both possible directions independently.

Of course obtaining the correct projectivity depends on the correct choice of correspondence. We have explored several heuristics to reduce the number of possible correspondences between pairs of sections from each image. Firstly, only a fixed number of the strongest pathways are selected from either view. Secondly, only a fixed number of the longest straight sections from these pathways are selected. Even with such heuristics it may not be possible to avoid searching through several possible correspondences. We have not explored mechanisms for 'evaluating' projectivities computed from candidate correspondences in order to select the most plausible.

Finally, figure 7 illustrates the projectivity produced from the two correct line correspondences shown in figure 6 by simulating the projection of the ground plane from the right view using intensity information extracted from the left image. The dark bushes and path boundary posts visible in the background and foreground respectively of the left image can be seen reversed in depth in the reconstruction.

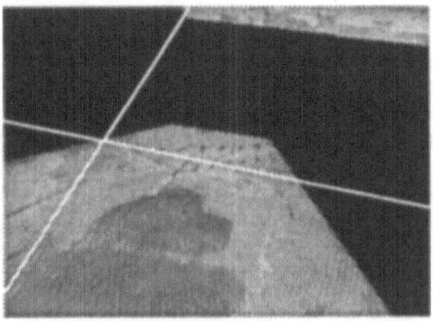

Figure 7: Simulation of the right image shown in figure 1 by reprojection of intensity information extracted from the left image.

7. Conclusion

There are many directions for further work to improve and extend the method developed above. More generally, statistical regularities extracted from long motion sequences provide a rich source of information about the moving objects themselves as well as the viewing geometry. The use of large sample sets can overcome some of the problems generally associated with the recovery of structure from images.

References

1. Duda R.O. and Hart P.E. "Pattern Recognition ans Scene Analysis". Wiley, New York, 1973.
2. Hogg D.C. "Model-based vision: a program to see a walking person". Image and Vision Computing, vol. 1, no. 1, pp.5-20, 1983.
3. Li-Qun X., Young D.Y. and Hogg D.C. "Building a Model of a Road Junction Using Moving Vehicle Information". in British Machine Vision Conference 1992, editor Hogg D.C. et al., pp.443-452, Springer-Verlag, London, 1992.
4. Mundy J.L. and Zisserman A. "Projective Geometry for Machine Vision". in Geometric Invariance in Computer Vision, editor Mundy J.L. et al., pp.463-519, MIT Press, Cambridge, 1992.
5. Nishihara H.K. "Practical real-time imaging stereo matcher". Optical Engineering, vol. 23, no. 5, pp.536-545, 1984.
6. Pollard S.B., Mayhew J.E.W. and Frisby J.P. "PMF: A stereo correspondence algorithm using a disparity gradient limit". Perception vol. 14, pp.449-470, 1985.

Mobile Robotics for the Surveillance of Fissile Materials Storage Areas: Sensors and Data Fusion

João G. M. Gonçalves, Gilberto Campos, Vítor Santos,
Vítor Sequeira, Filipe Silva

Commission of the European Communities,
Joint Research Centre,
Institute for Systems Engineering and Informatics
21020 Ispra (VA), ITALY

Abstract

A mobile robotics system is being developed to perform remote surveillance tasks inside a fissile material pilot storage area. The system is constituted of the operator's console and a remotely guided vehicle carrying a manipulator arm and sensors (i.e., odometers, TV cameras, ultrasound sensors and a laser range finder). Nuclear Safeguards philosophy stresses the need for having sensors based on different physical principles, allowing for true sensor independence and complementarity. These aspects are important when fusing sensor data, in order to build corroborative evidence upon which decisions are to be taken. The paper describes the mobile robotics system and studies the implications of the on-board sensors in the system's architecture both at the hardware and software levels. The use of sensor data for vehicle navigation is described, as well as the application of range images for environment authentication.

1. Introduction

The increasing amount of long term storage of fissile materials, specially those of strategic importance, in areas of difficult access, justifies the introduction of systems capable of remote verification tasks. International Nuclear Safeguards authorities verify the use of fissile materials throughout their complete life cycle. This activity stems from the application of the Nuclear Non-Proliferation Treaty. Sophisticated measurement systems (e.g., non destructive assays) have been put in place to help achieving those verification tasks.

These measurement systems record data taken at regular inventory campaigns. The continuity of knowledge between consecutive inventory campaigns is provided by surveying and confining the fissile materials. To be really effective, surveillance and confinement techniques must be based on a variety of sensors working upon different physical principles. This requirement for variety enhances overall sensor data redundancy, as well as, sensor data complementarity. Sensing devices must complement each other in the sense that they must contribute with enough corroborative evidence to build a consistent interpretation of the events that had been detected.

A mobile robot system is under development to perform remote surveillance tasks (e.g., visual inspection, reading of electronic seals, remote measurements) inside a fissile material pilot storage area. The main objective for the development of such a system is to demonstrate and evaluate the use of robotics technologies in this area of application. These technologies will be ultimately evaluated according to different factors, such as, functionality, performance, ease of use, robustness and abidance to the Nuclear Safeguards philosophy.

The main scope of this project is to have a working prototype capable of assessing the use of robotics technologies in this particular area of application. Several sensors working on different physical principles have been chosen. Their choice was based on the expected results that would be promptly available. Sensors, together with the computing systems that equip the mobile robot constitute the technological component of the project. As in most systems, the technology used deeply influences the design options that are to be taken. This includes the overall system's architecture both at the hardware and software levels. The result of this project must then identify for each sensor its utility in terms of the contribution to the performance of the overall system as defined by the functions that had been selected.

The paper describes the mobile robotics system being developed. The paper starts with a brief description of the project and the functionality that is sought. It continues with a detailed description of the sensors used, and their expected utility in terms of the required functionality. After presenting the system's architecture both at the hardware and software levels, the paper concentrates on the algorithms that were implemented for the different functions. The last section discusses the results achieved at the present stage of the project and introduces some ideas for future developments.

2. Project Description

Developments in computing and electronics technologies have widened the options available for designing a fissile material storage facility. Access to the storage areas, verification tasks (including inventory) and surveillance techniques must be reassessed to take into account the new available technologies. The SAOV project (Advanced Storage for Valuable Objects) [9,10] aims at building a pilot storage facility for nuclear materials, including remote surveillance and verification.

The remote verification system consists of:

i)   a vehicle carrying on board a manipulator arm, sensors (e.g., TV cameras, laser range finder, ultrasound) ·and associated equipment (e.g., pan and tilt units, motorised zoom and focus lenses);

ii) the system's operator console located outside the storage area.

The objectives of the remote verification system are threefold:

i)   to build a prototype system capable of demonstrating and assessing robotics technologies for the remote verification and handling of Nuclear Safeguards equipment;

ii) to provide a level of security and safety abiding to the Nuclear Safeguards philosophy and regulations;

iii) to provide quantitative data for high level material management, according to the concept that underlines an integrated safeguards system.

Before addressing the level of functionality that is required, some words are needed in what concerns the working environment. The nuclear material storage area is assumed to be a structured environment, i.e., the storage premises can be described by a CAD model. There is *a priori* knowledge in what concerns the dimensions and the topology of the storage premises and its contents (e.g., cupboards, lockers or storage cells), as well as their dimensions and locations.

The following operational tests reflecting routine work were devised:

i)   general purpose navigation;

ii) remote identification of containers using visual information, image processing techniques or bar code readers;

iii) remote verification of security seals;

iv) open cupboards, lockers and storage cells.

2.1   System's Functions

To accomplish these operations, the remote verification system must provide the following functions:

Vehicle navigation: This function comprises all tasks necessary for vehicle steering and manœuvring, including the corresponding human-computer interface, as well as the provision of navigational maps. Also included are the tasks for position monitoring and calibration.

Sensorial feedback: A decision that has been taken regarding the human-computer interface, is to have all sensorial information (i.e., TV images, odometric, ultrasonic and range data) available to the system's operator. Some of this data may be directly available (e.g., TV images), some other on request. This function is very much dependent on the type and quantity of sensors installed aboard the vehicle.

Manipulator arm control: All tasks that interface the manipulator arm contribute to this function. This includes the human-computer interface, as well as a real-time inverse kinematics module. This latter module works under close scrutiny from a workspace monitoring module in order to prevent collisions.

Provision of Safeguards related information: This function aims at making sure that Safeguards regulations and philosophy are observed. It is somehow an open issue, since it depends on the type and extension of controls that are implemented. At the moment, only an environment authentication module is being developed.

A major concern of the project is the safety of the mobile robot and its equipment. This aspect is paramount for the current application. The mobile robot must be protected from accidents, and its return to a parking station must be accomplished in safe conditions. The protection of the mobile robot also means the indirect protection of the storage premises and its contents.

Though safety aspects do not correspond explicitly to a system's function, it is clear that there should be mechanisms preventing the vehicle from losing control or causing damage to the environment.

2.2 Development Phases

The project is divided in three development phases:

Manual: All operations are directly controlled by the system's operator.

Semi-Automatic: The operator can interactively set some intermediate steps, e.g., tell the vehicle to go from one position to the next with the option of setting some points in between.

Automatic: Actions are objective driven, e.g., tell the vehicle to go to a particular place. Strategy planning modules and task schedulers are required.

The current status of the project places itself at the end of the first phase, with some work already done from the second phase.

3. Sensors

The following sensors are installed on board of the mobile robot:

Odometers: The vehicle has two motorised rear wheels and two free wheels in the front. Both rear wheels have incremental optical encoders installed on their axis. The manipulator arm has one incremental encoder at each of its six joints. These encoders and their associated counters constitute the basis for the odometers. Odometric data is read by a VME multi-axis controller board and is available to all processes that request it.

TV Cameras: Two TV cameras, providing standard CCIR 50Hz video signal are mounted aboard the vehicle. Both cameras are mounted on top of two computer controlled pan and tilt units, and are equipped with computer controlled zoom and focus motorised lenses.

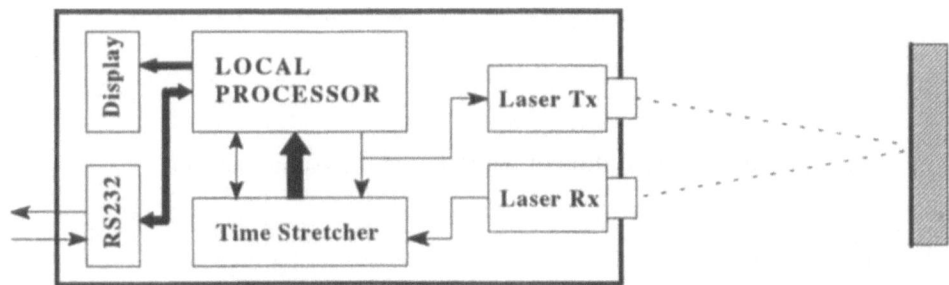

Fig. 1: Description of the laser range finder.

Laser Range Finder: The laser range finder (LRF) works on the time-of-flight principle, i.e., the distance between the instrument and the target is obtained by measuring the delay between the time at which a laser pulse is sent and the time the corresponding echo is received (Figure 1). The LRF is interfaced by means of a standard serial line, transmitting commands from a host, and receiving the measurements in ASCII format. The LRF operates on a range of distances between 1 and 20 metres. The laser beam has a diameter of about 4 cm and a beam divergence of 0.16 degrees. The LRF emits laser pulses at a fixed rate of 4 KHz, and averages the echo delays during a fixed programmable time T varying from a minimum of 25 ms to a maximum of 5 seconds. Distance resolution is about 0.8 cm, and the precision varies with the measurement time from 25 mm for T=25 ms to about 6 mm for T=5s [2,5].

Ultrasonic Sensors: A belt of 24 ultrasonic sensors of the Polaroid type has been mounted around the vehicle. Sensors are grouped in nodes, each node being separately activated. When activated, each sensor emits an ultrasonic pulse inside its cone of influence with an aperture of about 20 degrees. Ultrasonic pulses are reflected by surrounding objects located within its cone of influence, and the echoes measured by special purpose electronic circuitry (Figure 2). Ultrasonic sensors operate within a range of distances between 15 cm and 10 metres [16]. The ultrasonic sensors are driven by special hardware (a VME board inside the mobile robot's computer cage).

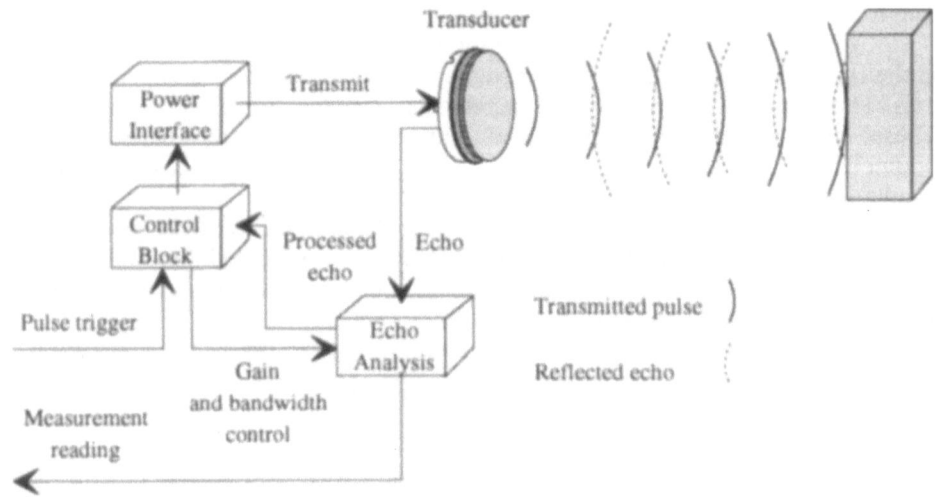

Fig. 2: Description of the Polaroid ultrasonic sensors.

System's Tasks

The functions mentioned above are not independent, in the sense that they share and require common data. From an architectural point of view it was decided to implement the whole system using modular and cooperative tasks. A given function, say, vehicle navigation, is built by making coordinated accesses to the relevant tasks. It is possible to classify the tasks in two categories:

i) *low-level*, i.e., hardware or sensor oriented, and

ii) *intermediate level*, i.e., application oriented.

Low-Level Tasks

a) *Communications Server*: manages the serial link between the system's operator console and the mobile robot. Included in the communication server are watchdogs for surveying the communications link and for bi-directional transmission of alarm messages, e.g., collision detection or emergency stop.

b) *Ultrasonic Sensors Server*: provides data from the ultrasonic sensors.

c) *Pan and Tilt Control*: interface for the on-board pan-and-tilt units for the orientation of the TV cameras and laser range finder.

d) *Range Finder Control*: interfaces the laser range finder, and includes all the commands required for the acquisition of range profiles and range images.

e) *Video System Control*: control and display of the images from the on-board TV cameras; it includes controls for image contrast and brightness, as well as for the motorised zoom and focus lenses.

f) *Emergency Detection*: set of high-priority watchdogs monitoring for conditions that could endanger the vehicle and environment, such as collision detection or communications failure.

Intermediate Level Tasks

g) *Steering and Manœuvring*: provides the operator with several ways of controlling the speed and orientation of the mobile robot, including a semi-automated trajectory simulation and generation capability.

h) *Navigation Maps*: 2D and 3D real-time animation of the mobile robot within the environment as "observed" from a user selected viewing point.

i) *Auto-Focusing*: provides an automatically focused image for any of the on-board TV cameras.

j) *Object Tracking*: automatically displays a focused image of a moving object with known position (e.g., the manipulator arm end-effector).

k) *Position Monitoring*: ascertains the consistency between the robot's real position measured by the sensors with the values provided by the odometers.

l) *Position Calibration*: computation of the exact position and orientation of the mobile robot inside the storage area.

m) *Depth Maps*: provides 3D data from the environment.

n) *Environment Authentication*: certifies the position of the mobile robot inside the storage premises, based on data measured by the on-board sensors and *a priori* knowledge of the environment.

o) *Manipulator User's Interface*: interface for controlling the manipulator arm.

p) *Inverse Kinematics*: transforms operator's commands for the displacement of the manipulator arm end-effector into manipulator position commands.

q) *Workspace Monitoring*: monitors the displacements of the manipulator arm, and alerts the operator for situations where workspace limits or collisions are bound to occur.

	Vehicle Navigation	Sensorial Feedback	Manipulator Arm Control	Provision of Safeguards Data
Low Level Tasks				
Comm. Server	✓	✓	✓	✓
Ultra Sound Server	✓	✓		
Pan and Tilt Control	✓	✓	✓	✓
Range Finder Control	✓			✓
Video System Control	✓	✓	✓	✓
Emergency Detection	✓	✓		
Intermediate Level Tasks				
Steering and Manœuvring	✓			
Navigation Maps	✓	✓	✓	
Auto Focusing	✓		✓	
Object Tracking	✓		✓	
Position Monitoring	✓			
Position Calibration	✓			✓
Depth Maps				✓
Environment Authentication				✓
Manipulator User's Interface			✓	
Inverse Kinematics			✓	
Workspace Monitoring			✓	

Table 1: Dependencies between functional capabilities and system's tasks.

Table 1 lists the dependencies of system's functions on low and intermediate levels tasks. Table 2, below, lists the dependencies of intermediate level (i.e., application oriented) system's tasks on the installed sensors. It can be seen that not all the tasks are directly or indirectly dependent on all the sensors. More detail on how each task uses the different sensors is given in further sections.

	Odometers	TV Cameras	Ultrasound Sensors	Range Finder
Steering and Manœuvring	✓	Manual	Research work with Neural Networks	-
Navigation Maps	✓	-	Research work with Neural Networks	-
Auto Focusing	-	✓	-	-
Object Tracking	✓	-	-	-
Position Monitoring	✓	-	✓	-
Position Calibration	✓	-	-	✓
Depth Maps	✓	-	-	✓
Environment Authentication	✓	-	-	✓
Manip. User's Interface	✓	✓	-	-
Inverse Kinematics	✓	-	-	-
Workspace Monitoring	✓	-	-	-

Table 2: Utilisation of sensors in the different tasks.

4. System's Architecture

A representation of the system's hardware architecture is shown in Figure 3. Commands between the host computer and the mobile robot's computer are transmitted via an RS232 connection (cable or wireless link). These commands are managed by the Communication Server running on the robot's computer and redirected to the appropriated real-time task. The tasks in the robot are executed concurrently under Albatros, a real-time multitasking operating system.

Background tasks both in the robot and in the host computer, monitor the status of the communications between the robot and the host. If there is a

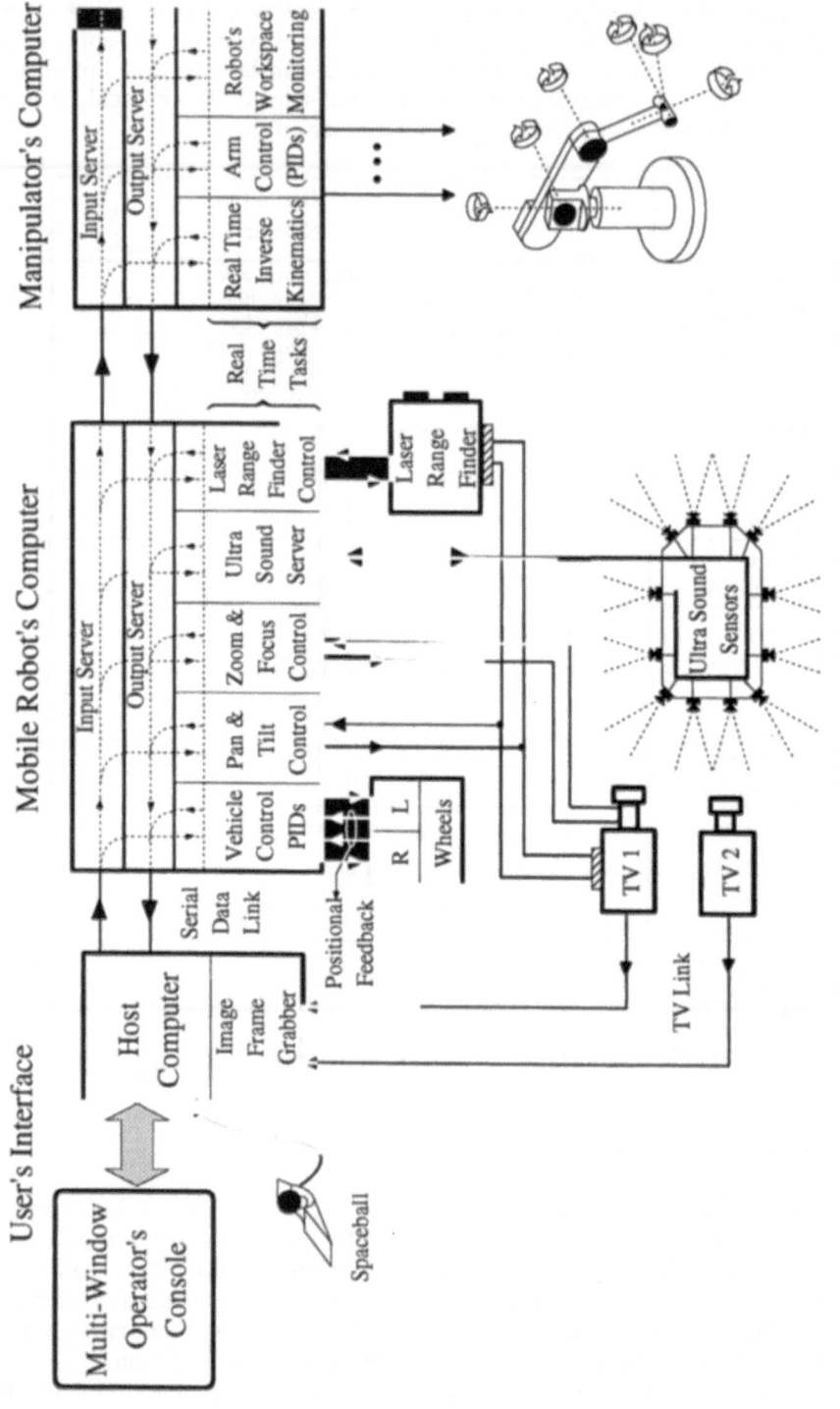

Fig. 3: System's Hardware Architecture.

communication failure the robot stops immediately, enters a standby state, and the operator is informed.

It can be seen from Figure 3 that apart from the host computer, the mobile robot has its own computers. This configuration allows for the distribution of intelligence amid the available computers. Three criteria have been considered in what concerns the decision of where to run a given task:

i) to have all application oriented tasks, including human-computer interface, running in the host computer, and all hardware oriented tasks running on one of the computers aboard the mobile robot;

ii) to minimise the amount of information through the serial line. Ideally, only high level commands or measurements should occupy the serial line;

iii) to minimise the time delay introduced by a task in an interactive control loop, i.e., a control loop in which the operator is involved (e.g., operator command, action, control, sensor's feedback, new command, etc.).

The combination of the above principles, though not determinant, reduces the number of available options, and hence conditions the allocation of a computer to a particular task.

5. Software Architecture

The system's software architecture is represented in Figure 4. Some of the system's tasks (see Section 2.4) are executed in the operator's workstation, and the remaining in the mobile robot's computers. For the sake of clarity, and without loss of generality only one remote computer is represented.

All tasks are modular in the sense that they do not directly depend on other tasks and can run on a stand-alone basis. A key task for this software architecture is the Communication Server, which has been designed with the following objectives:

i) to provide modularity at the system level, by creating a "standard" for process communications;

ii) to provide the system's intrinsic data, i.e., system's hardware configuration, environment's CAD model, sensors' data, and system's operational status;

iii) to code, interpret and filter commands to and fro the mobile robot;

iv) to rationalise the serial communications traffic.

The Communication Server consists of several tasks, one running on the workstation, and the remainder on the remote computers. On the workstation, the different tasks exchange data with the Communications Server by means of the Unix socket-based inter-process communications. On the remote computers, communications between tasks use shared memory mechanisms.

All messages to and fro the workstation pass through the Communications Server, which is closely coupled to the Emergency Detection task. The reason lies in the need for prompt action in case of emergency. This close link facilitates speed, since it allows the Emergency Detection task to check sensors' data and commands in first hand. Special high-priority software emergency links (Figure 4) were created to stop all mobile robot's activity, i.e., vehicle's motion and displacements of the manipulator arm, when an emergency occurs.

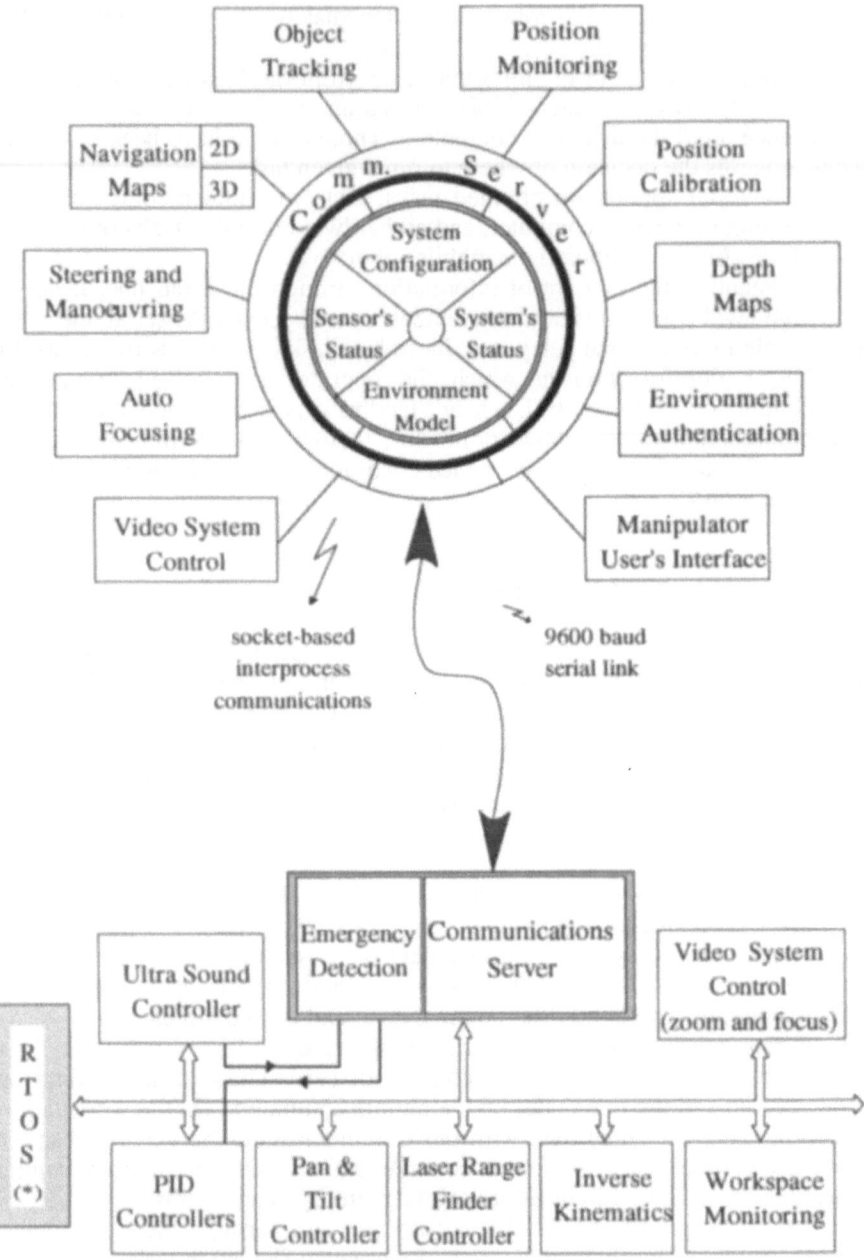

Fig. 4: System's Software Architecture

6. Human Computer Interface

The system's console is a graphical workstation from which the system's operator controls and monitors the mobile robot. Considering that most of the acceptance of the project will depend on the judgement of potential users, great effort was dedicated to the development of the human computer interface.

The main characteristic of the interface is the integration into a single screen, of all the information needed by the system's operator. This is considered a major change in what concerns "traditional" robotics and tele-operation interfaces with large consoles, many screens, buttons, knobs, etc. This is an important aspect since an integrated display requires a more active participation from the operator, rather than just watching the events evolve in a multi-screen console. Details on the human computer interface can be found elsewhere [13, 14].

Figure 5 shows a typical screen of the operator's console. The following tools can be seen: Steering and Manœuvring, TV cameras display (2 cameras), 2D and 3D Navigation Maps, and the Ultra-Sound Sensor Display. It is up to the operator to select at any particular moment which are the tools he or she considers relevant for the current operation.

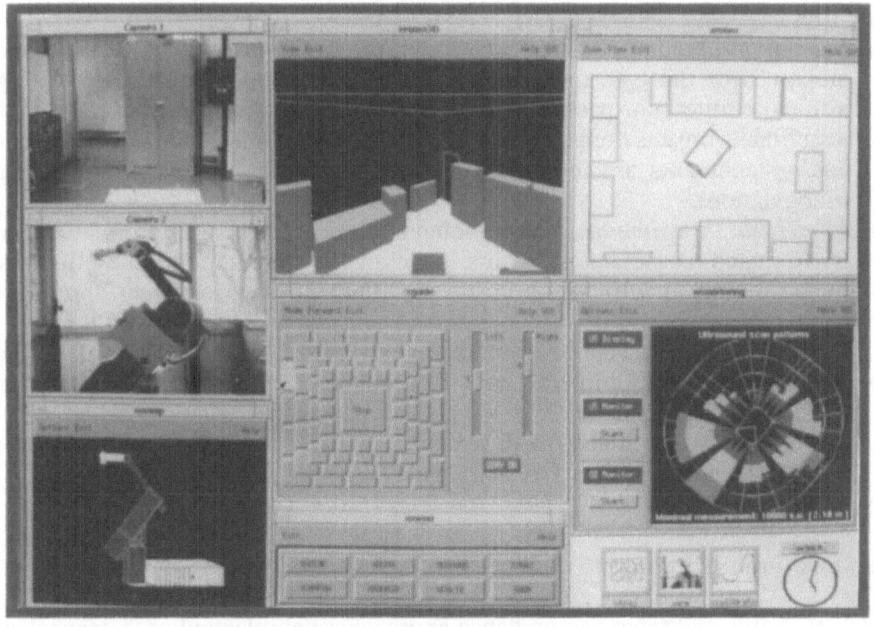

Fig. 5: Mobile Robot's Human Computer Interface.

7. Robot's Localisation

A major concern in mobile robotics is to have at each moment a good estimate of the position of the mobile robot (x, y, θ). Navigation procedures, whether they are manual or automatic, assume an accurate estimation of current position coordinates to generate motion commands. Several approaches for position estimation can be used, most of them based on the interpretation of positional sensors' readings. The two basic approaches are [4]:

i) *Dead-Reckoning*: the position coordinates refer to a reference coordinate system fixed to the mobile robot. Odometry, a simple, fast and totally self-contained method, is normally used. Odometry, however, is rather unreliable since it drifts with time, due to wheel slippage or surface roughness, generating time cumulative errors. There is then the need for regular calibration of the odometers.

ii) *Reference Guidance*: the position coordinates refer to a reference coordinate system external to the mobile robot. On-board sensors provide measurements relating the robot with the environment, such as: distances to the environment, angles between *a priori* known landmarks in the environment (e.g., radio beacons), or time differences between emitted signals (e.g., satellite systems).

A major difficulty is to extract the position coordinates from the measurements, considering that the overall precision depends on how accurate the measurements are, or on the exact detection of landmarks. Several methods using reference guidance have been described. These methods fall generally in two basic categories:

a) *Beacon and landmark based estimators*: the environment must be previously prepared for the type of sensors used. Beacons can be either active (e.g., infrared emitters) or passive (e.g., mirrors), and landmarks are normally well identifiable objects (e.g., bar codes, concentric circles). Once the beacons and/or landmarks are found, the robot's position can be determined (e.g., triangulation).

b) *Model based estimators*: these methods work upon range information, and try to match measurements with a model of the environment. These methods do no require modifications in the environment. Two different approaches can be found:

- *feature based*: a set of features are extracted from the sensors' data (e.g., line segments, corners) and then matched against the corresponding features in the model.
- *iconic*: works directly with sensors' raw data, minimising the discrepancy between sensed data and an *a priori* known environment model.

7.1 Positioning Strategy

Dead reckoning and reference guidance approaches complement each other. Dead reckoning provides fast position estimates. Reference guidance, instead, is slower though more accurate in the long term. It is normal practice to have navigation procedures based on dead reckoning and regularly calibrate the mobile robot's position coordinates with a reference guidance method. This practice was followed in the present project.

The choice of a localisation method depends basically on the application. Some solutions are possible in one application area and less feasible in others. Many widely used commercially available localisation systems require the installation of beacons or landmarks in the environment. In the present case, the need for predisposing the environment to the navigation system was considered a disadvantage, for the following reasons:

 i) it is not practical, since it requires the consent from the plant operator;
 ii) it is not considered a general solution, since it depends on a particular technology and environment predisposition, and last but not the least,
 iii) it may be considered as a dependence in what security is concerned.

It was decided to have the Position Calibration task based on data from the on-board laser range finder and use a model-based position estimation method. Consequently, the positioning tasks are as follows: on one side are all navigation related tasks, such as Steering and Manœuvring, 2D and 3D Navigation Maps, working upon data measured from the odometers; on the other side is the Position Calibration task, working with the distance profiles measured by the range finder.

An implementation problem occurs on how to determine when the odometers need calibration. Indeed, there is no "rule" saying that after a given amount of time odometry data are unreliable. Odometry precision depends strongly on the sort of ground surface the robot is moving on, and on the number and type of manœuvres since the previous calibration. This problem may seem irrelevant from a theoretical point of view, but it is not from a practical perspective. Our range finder is a heavy piece of equipment (approximately 3.5 kg), and it takes some time to acquire the distance profiles. It is therefore recommended to stop the vehicle while calibrating the position.

To overcome this difficulty, and reduce the number of unnecessary calibrations, it was decided to have a position monitoring task. The aim of this task is to ascertain the consistency between odometric data and the robot's real position as measured by the on-board sensors. Considering that the laser range finder is a "slow" device, the Position Monitoring task will be based on data from the ultrasonic sensors.

7.2 Position Monitoring

The position monitoring task runs normally as a background task, since there is no need for interaction. The task must not interfere with normal operation, though its results should be known in real-time. The task includes two position monitoring tests:

 i) Geometric Inconsistencies, and
 ii) Ultrasound Data Analysis.

Geometric Inconsistencies Test

The geometric inconsistencies test, checks for logical inconsistencies between odometers' readings and the environment. Figure 6 shows a non-logical position. The basic test is to check for intersections between the robot's and the environment's representations. If such an intersection is detected, the robot must be calibrated.

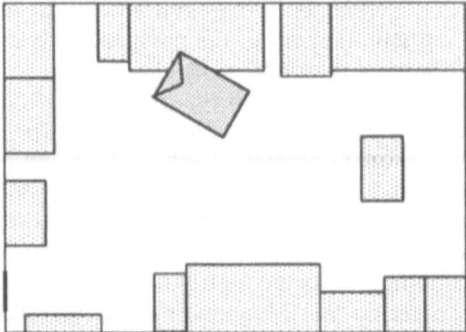

Fig. 6: Geometric Inconsistency Test: impossible robot position.

The algorithm tests the intersection of the polygon representing the mobile robot, and all polygons representing the environment. Polygon intersection is detected whenever a vertex of the robot's representation is inside one of the polygons representing the environment. To check whether a vertex is inside a polygon the following distance measurement is used:

$$d(x_0, y_0) = \frac{ax_0 + by_0 + c}{\left(a^2 + b^2\right)^{\frac{1}{2}}}$$

(Eq. 1)

where x_0, y_0 represent the vertices' coordinates and a, b, c represent the parameters of one of the line segments that constitutes the polygon. The sign of d indicates on which side of the line the point is and the magnitude represents the distance between the point and the line. A polygon P is represented as an ordered list of vertices, as follows:

$$P = \left\{ \left(x_k, y_k\right): \ 1 \leq k \leq n+1 \right\}$$

(Eq. 2)

where the n vertices of P are ordered counter clockwise about the centroid. To detect whether or not point (x_0, y_0) lies inside polygon P, the distance function $d(x_0, y_0)$ is successively computed using the lines generated by the edges of P. The point (x_0, y_0) lies inside P if and only if the n distances $\{d_k(x_0, y_0)\}$ are all positive. If point (x_0, y_0) is inside P, the amount of penetration is equal to the minimum value of the n distances. It is important to note that this method is valid only for convex polygons, though that there is no loss of generality, since non convex polygons can always be regarded as union of convex polygons. It is worth noting that the primitives used to describe the objects always represent convex shapes.

Ultrasound Data Analysis

Figure 5 shows the ultrasound data display tool. There are 24 cones around the robot, each one representing an ultrasound sensor. The length of each cone is proportional to the distance measured by the corresponding sensor. For some cones their area of influence overlaps. This tool provides a dynamic representation of the distances measured by the different sensors. It can be seen that the outline shape of the different ultrasound cones matches the environment outline, as seen in the 2D Navigation Map. The ultra sound display tool helps the operator to prevent collisions and may indirectly determine the need for position calibration.

Current work is in progress for the automatic analysis of the ultra sound sensors data. Position monitoring checks for consistency between the location given by the odometers and the distances from the ultra sound sensors to the environment. The information given by the ultra sound sensors combined with data from the odometers and the *a priori* knowledge of the environment topology may indicate the need for calibration.

7.3 Position Calibration

The Position Calibration task computes the exact position and orientation of the robot inside the storage premises. The exact position is then used to recalibrate the odometers and indirectly all navigation tools. The method used is totally self-contained. It is based on an algorithm described by Cox [4] and does not require any environment modification.

Position estimation is based on data from two sensors: range finder and odometers. The odometers provide an initial position estimation and the range finder provides accurate data to correct the initial estimation. It is worth noting that the algorithm still works, though slower, with no initial position estimation. The algorithm requires the following information:
  i)   the environment representation;
  ii) the current position given by the odometers, and
  iii) a horizontal profile of the distances between the robot and the environment (measured by the range finder).
A matching algorithm registers the range data with the environment representation and determines the correct position of the robot. In case of ambiguity, e.g., symmetrical environments, the initial estimation provided by the odometers determines the correct answer.

Figure 7 shows the results of the Position Calibration task. The continuous line represents the environment's map, whereas the small rectangle represents the location of the robot provided by the odometers. The dotted line represents the distances between the robot and the environment measured by the laser range finder. Each dot is a distance measurement for a particular orientation. Distance measurements were taken at regularly spaced angles. It is worth of note that dot density is bigger for short distances, and smaller for larger distances. In the picture on the left, distances are plotted taking as reference the position of the robot provided by the odometers. The Position Calibration task attempts to match the dotted profile with the *a priori* known environment map. The results are shown in the picture on the right. The distance profile is plotted taking as reference the position of the robot given by the algorithm. The two lines, i.e., the environment map and the distance profile, overlap. Some jitter can be noticed. This is due to errors while measuring the distances (i.e., noise in the laser range finder and scanning imprecision), and to the fact that the environment map is a rough approximation of reality. Indeed, in a prefabricated building walls cannot be modelled as straight perpendicular lines, and there are items that had not been completely modelled.

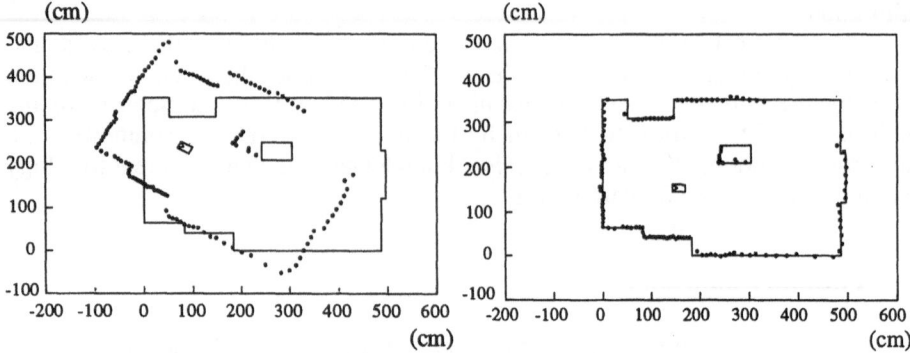

Fig. 7: Position Calibration: robot's position before and after calibration.

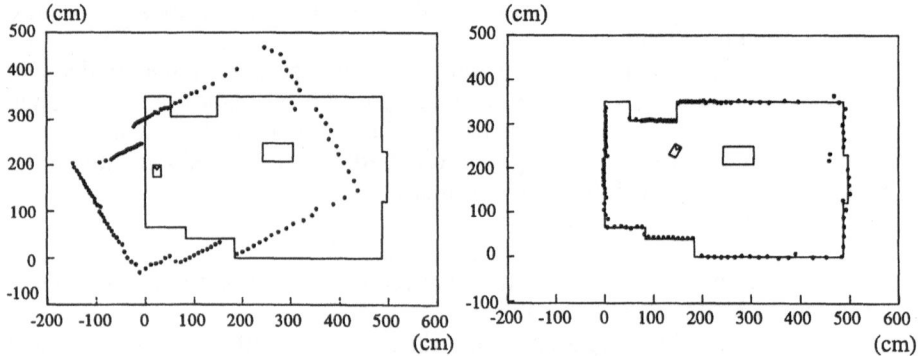

Fig. 8: Position Calibration: robustness test.

The algorithm is quite accurate, and the final position precision is better than 2 cm. Several tests were made to check the algorithms' robustness. Some modifications to the environment were made, without updating the environment representation, e.g., objects were added or removed. The algorithm worked well, and in all situations a convergence point was reached, matching the correct robot's position. Figure 8 shows the results of such a test.

8. Environment Authentication

A characteristic particular to this area of application is the requirement for having sensors based on different physical principles. Each sensor contributes with its own data in order to add corroborative evidence to the remote verification task.

The environment authentication tool must then be able to provide an answer to the following question:

"Considering that the mobile robot is supposedly located at position (x,y,θ) how does the environment information measured by the sensors match the *a priori* known environment model?"

Considering the characteristics of the on-board sensors and the fact that visual feedback from the TV cameras is always expected to be available, the laser range finder was selected to provide the extra information to verify and authenticate the environment. It is worth noting that the above question does not correspond to a recognition problem, but rather to a validation or authentication one.

Different approaches have been pursued, all of them working upon range images. Distance information has characteristics that make it very difficult to reproduce, unless the exact geometric conditions are known. Furthermore, distance information as measured by the laser range finder does not depend on a predisposition of the environment, e.g., illumination conditions.

8.1 Range Image Acquisition

Generally speaking, an image is a matrix with measurements of the spatial distribution of a given property. For range images, this property is the distance to a reference, normally a plane or a point. In the present case, the reference is a point located on the laser range finder (LRF). Alternatives are described elsewhere [8].

By making the LRF scan along the X and Y orientation axis, it is possible to build the raw matrix with the range information as represented in Figure 9. It should be noted that range data should be always associated to the intrinsic geometry of the scanning mechanism. Indeed, range matrices will be different if scanning mechanisms with different geometries are used.

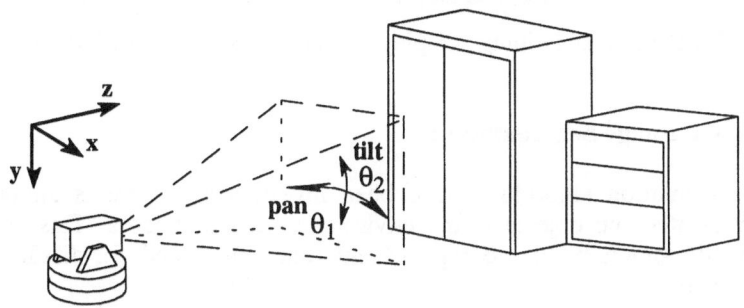

Fig. 9: Range image scan.

Figure 10 a) shows a range image from an office scene. The office layout is shown in Figure 10 b). There are some points worth notice:
a) the image has 128x128 pixels (rangels in this case). Each pixel represents a distance ranging from 1 to about 4 meters. The image corresponds to an environment scan with 60 degrees aperture both in the horizontal and vertical axis. Measurements were made at angles equally spaced in both directions.

b) The image shown in Figure 10 a) is a visual representation of the range image, with a particular grey level map: shorter distances are lighter, and longer distances are darker. Other grey level maps could be possible.

c) It can be seen that some straight lines, e.g., the corner between the wall and the ceiling, do not look straight in the image, but curved. This effect is due to the fact that the scanning mechanism had a spherical geometry and the image is being displayed on a plane.

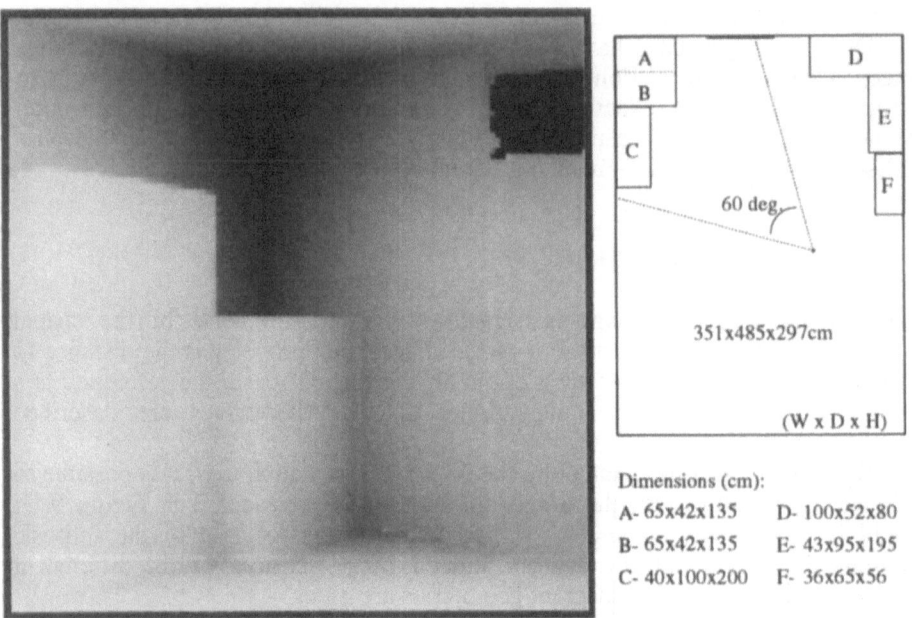

Fig. 10: a) Range image display of an office scene; b) office layout.

8.2 Range Image Segmentation

Two segmentation algorithms have been investigated. There is an underlying assumption that the objects in the environment are polyhedra. This assumption corresponds generally to the type of environments found for fissile materials storage areas.

i) *Edge Detection*: Edges are searched in the image, and regions can be extracted as the image areas delimited by the edges.

ii) *Planar Region Extraction*: The image is divided into a set of non overlapping planar regions. Visible object edges can be then identified as the intersection of adjacent regions.

These two approaches are dual in terms of the geometric properties that are being looked for. The underlying algorithms also work on different principles, in the sense that one is based on local operators, and the other on a global geometric property.

Edge Detection

An edge detection tool was implemented based on the algorithm described by Al-Huzaji and Sood [1]. The algorithm works in two parts. The first part consists of the detection of the step edges in the image. The second part addresses the detection of roof edges. The algorithm is based on a local convolution kernel using residual calculations to detect if a given pixel belongs or not to an edge. Figure 11 shows the results obtained for a range image.

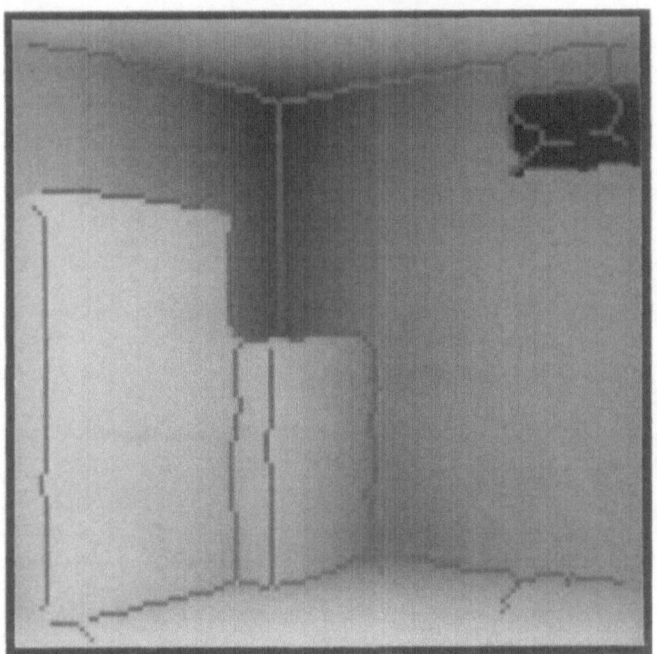

Fig. 11: Edge extraction of a range image.

Region Detection

The region detection algorithm divides the image in non overlapping planar regions. It is based on a modified version of the split and merge algorithm described by Taylor et al. [15].

Since planarity is a global property, there is the need to make a geometrical transformation to the image before applying the region detection algorithm. The initial coordinates of each pixel $(\theta_1, \theta_2, R(\theta_1, \theta_2))$ are transformed into the correspondent (x, y, z) Cartesian coordinates. Included in this transformation is also the correction for the geometry of the scanning mechanism.

A new image is generated by computing for each pixel the best plane fitting the pixel's neighbourhood. This last image is then encoded (split phase) into a quadtree, using as homogeneity criterion the three plane parameters. Adjacent patches belonging to the same planar region are then merged (merge phase) into larger regions. The merge phase is iterative to allow for larger regions.

Figure 12 shows the results of the segmentation. Different planar regions are identified, such as, walls, cupboard doors and side, file cabinet front and side. More details on the algorithm can be found elsewhere [3].

Fig. 12: Segmentation of a range image of an office scene into planar regions.

8.3 Scene Matching with Range Occupancy Grids

Another approach for environment authentication is based on range occupancy grids. This method creates an occupancy grid (i.e., tridimensional matrix) located at the mobile robot's supposedly known position (x,y,θ) (Figure 13).

In the first step, the algorithm fills (i.e., sets the value to 1) those cells occupied by objects as described by the environment's CAD model. In the next step, for each image pixel, the algorithm computes the coordinates (x,y,z) of the corresponding grid cell. If the grid cell is filled, there is match between the distance measurement corresponding to that particular pixel and the environment. If the grid cell is empty, there is a mismatch.

It should be noted that this method assumes the objects to be hollow entities, rather than solid. The reason for doing this lies in the fact that the only distances available are the ones till the objects' surfaces. If the objects were not considered hollow, then the number of matches would wrongly increase whenever a position error caused the grid to be closer to a wall or object than actually foreseen by the robot's position (x,y,θ).

A match index is computed as the ratio between the number of matches and the number of measurements, i.e., the number of pixels in the range image that fall inside the grid. It is worth notice that there is no requirement for all filled in grid cells to be accessed. Indeed, accesses to grid cells depends strongly on how distant the cells are from the robot and on the spatial (i.e., in the present case, angular) resolution that characterised the acquisition of the range image.

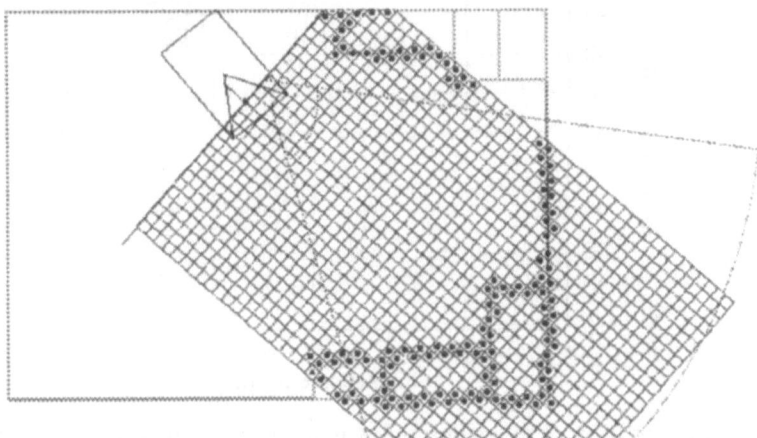

Fig. 13: Range Occupancy Grids: the grid has its origin at the mobile robot's position and grid cells are filled with the expected view of the environment.

Figure 14 shows the variation of the match index with grid cell size for two office scenes. The match index generally increases with grid cell size, and beyond a cell size of 7.5 to 10 cm the match index is greater than 90%.

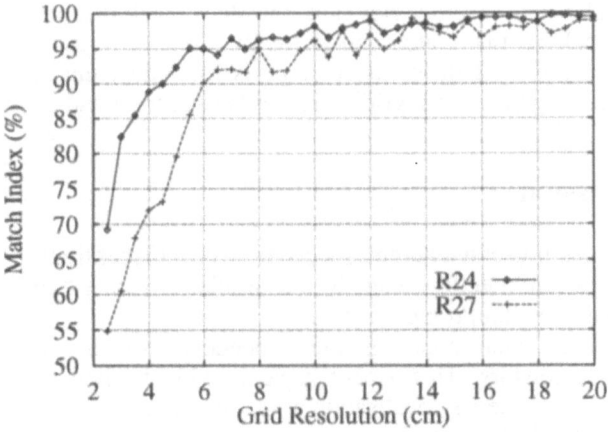

Fig. 14: Variation of the match index with the grid cell size.

Figure 15 shows the variation of the match index with the orientation error, for different grid cell sizes. It can be seen that the method is very sensitive to orientation errors. In fact, for orientation errors larger than, say, 3 degrees the match index falls steeply.

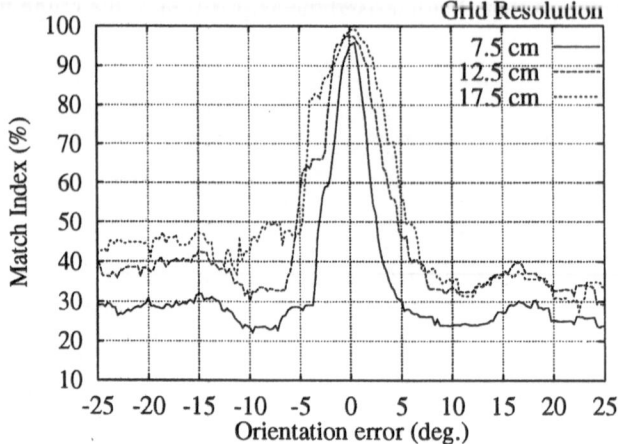

Fig. 15: Variation of the match index with the orientation angle error.

Figure 16 shows a set of plots representing the variation of the match index with a position error (Δx,Δy) with a null orientation error, for various grid cell sizes. It can be seen that the match index has its peak for a null position error. It then decreases with the position error magnitude, though some local ridges are seen for particular directions of the position error vector. It must be noted that these directions are environment or context dependent, in the sense that different environments or different "viewing" points produce different match index plots.

Another relevant finding is that there is no much interest in using larger grid cell sizes. The plot is rather flat (i.e., the sharp peak disappeared) for a cell size of 20 cm making the match index totally useless, without any discriminating ability. An analysis of different error plots from different scenes contributed to establish the range of useful grid cell sizes between 7.5 and 10 cm.

9. Discussion

From the work presented some aspects are worth discussion:
  i) System's Architecture, namely the problem of distributing the tasks among the available computing systems;
 ii) Sensors in what concerns the dependency of the whole system on the available sensors. This aspect became evident when there was the need to design the system's localisation tool, required to provide fast and accurate results simultaneously.
iii) Range images and the methods developed for image authentication.

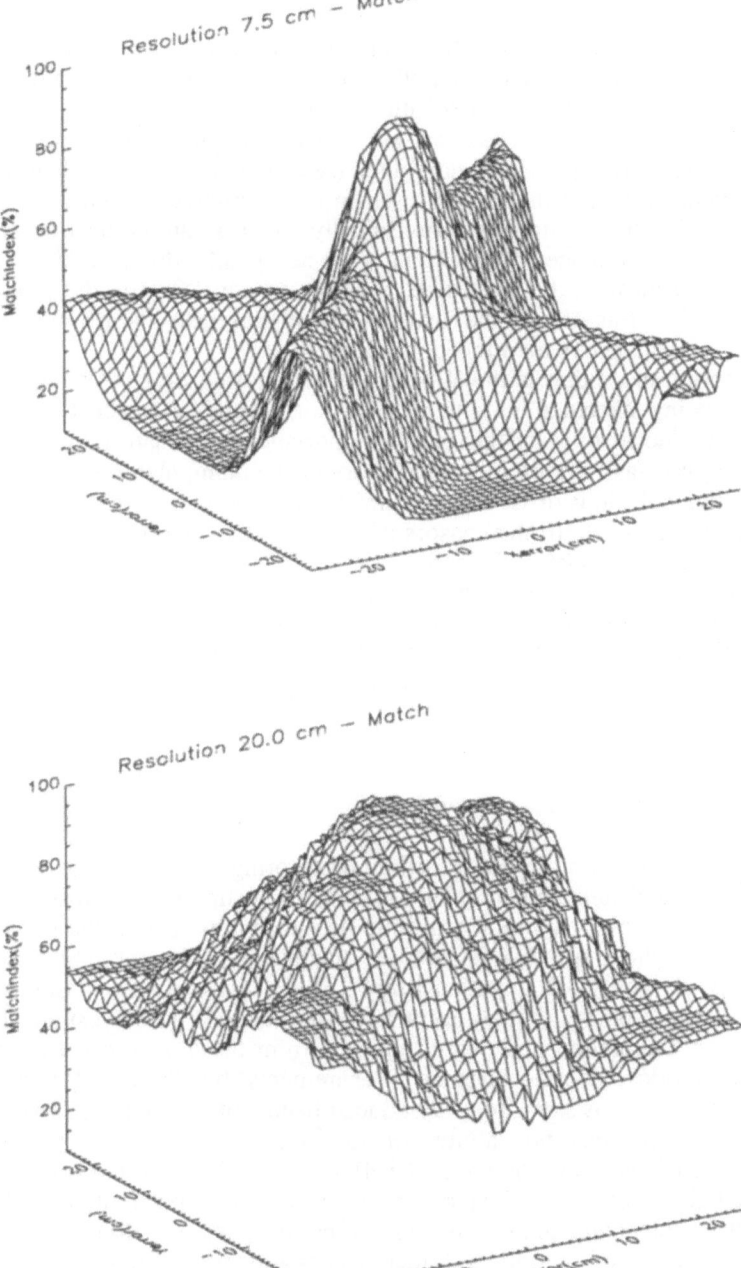

Fig. 16: Variation of the match index with the position error.

9.1 Distribution of Tasks amid the System's Computers

Generally speaking, all sensor or hardware oriented (i.e., low level) tasks are executed on the robot's computer and all application oriented are executed at the workstation level. There are two exceptions to this rule. The first exception can be found for the Inverse Kinematics task, which is application oriented and is executed on the manipulator arm computer (see Figures 3 and 4). This task is computer intensive and is thoroughly used in an interactive procedure, i.e., the operator control of the manipulator arm. Initially, the task ran on the workstation, but the overall performance was far from being acceptable, due to the amount of traffic generated through the serial line, and consequent delays in the interactive control loop. The transfer of the Inverse Kinematics task to the manipulator computer solved the above mentioned problems.

The second exception is the Workspace Monitoring task. The reason why this task is being executed on the manipulator computer rather than on the workstation is due to the fact this task is important for the safety of the mobile robot. Indeed, this task evaluates the trajectory of the manipulator arm in order to predict possible collisions or workspace limits. The character of this task makes it desirable to work closely to the sensors data, and to the Inverse Kinematics task for prompt inhibition if such is the case.

A task that may seem like an exception is the Video System Control task, which is sensor oriented and is executed at the workstation level. A look at Figure 3 shows that the TV cameras are connected to the workstation rather than to the mobile robot. It is then clear the reason why this task is executed on the workstation.

9.2 Hierarchy for Position-related Tasks

As explained in Section 7, the Position Monitoring task was introduced to overcome the implementation problem of when to calibrate the position of the mobile robot. The solution adopted introduces a hierarchy among the positioning tasks. At the highest speed level are the navigation tasks (i.e., Steering and Manœuvring, 2D and 3D Navigation Maps). These tasks rely on odometric data which are fast to retrieve, though not always accurate. At the lowest speed level is the Position Calibration task, based on data from the laser range finder. This instrument provides accurate distance measurements, but is limited in terms of speed. These limitations stem from its reduced bandwidth (about 20Hz) and from its weight, which does not allow for fast scan movements. The Position Monitoring task (intermediate speed level) works in background, checks how uncalibrated the robot is, and alerts the operator when position discrepancies exceed a given threshold, and prompts him or her for position calibration.

It should be noted that this task hierarchy is a mere consequence of the type of sensors installed aboard the mobile robot. Indeed, had the sensors been different or had they had different characteristics (e.g., speed), the system's software architecture and task division would certainly be different. It is rather unlikely that navigation tasks cease their dependence on odometric data. It is, however, probable that the ultrasound sensors would be no longer needed, had the range finder a larger bandwidth, or was it lighter (i.e., allowing for faster scans). The existence of the Position Monitoring task is only meaningful as long as the time for position calibration (i.e., data acquisition and processing) is long enough to require the vehicle to be still while calibrating.

9.3 Range Image Processing

There must be geometrical coherence along the processing chain, i.e., scanning mechanism, range acquisition parameters (i.e., equally spaced angles or Cartesian coordinates), processing algorithms, and analysis algorithms (i.e., local vs. global geometrical properties). All chain elements must work using a uniform geometric model, otherwise coarse processing and interpretation errors are bound to happen.

The range occupancy grid method differs from the ones based on image segmentation in the sense that the environment is directly evaluated against the range image without the need for the extraction of spatial features. Both approaches, i.e., range occupancy grids and image segmentation, are model-based, the former relying on statistics, and the latter on structural features (e.g., edges, or planar regions). One should not forget that the scope of this work is environment authentication, rather than environment modelling or recognition.

10. Future Work

Apart from the natural developments in the work described above, there are two areas on which attention will concentrate:
  i) Vehicle Navigation, and
  ii) 3D Environment Modelling.

10.1 Vehicle Navigation

It is believed that most of the tools developed so far, have reached a reasonable level of quality in terms of both functionality and user's interface. These tools are however limited to the manual operation of the mobile robot. There are nevertheless circumstances requiring autonomous vehicle control, even during the current phase of the project, i.e., the manual phase. Indeed, if communications between the operator's console and the robot do break down, stopping the vehicle is not enough. Mechanisms should exist to guide the mobile robot to a pre-defined parking area, where plant operators can retrieve it in an orderly manner.

Two different approaches will be investigated. The first one is based on the use of neural networks for processing ultrasonic data. Details on the work that was developed so far are given below.

The second one is based on the use of vision. In recent years, computer vision methods achieved a qualitative leap. Indeed, work on optical flow, stereoscopic vision (with two or more cameras), active vision, model-based vision, geometrical reasoning, etc., has provided some of the tools that can contribute to autonomous vision-based navigation. A word of caution is due however, in what concerns the computing requirements needed by a "real-time" vision-based navigation system.

Neural Networks for Vehicle Navigation

Ultrasonic sensors have good qualities in terms of cost, speed and precision in measuring distances. Ultrasonic sensors have however some disadvantages in what concerns their ability to locate obstacles. This becomes a serious drawback when a single sensor is used or when in the case of a multi-sensor system, their areas of influence do not overlap. Without this ability, ultrasonic sensors see their use restricted to obstacle detection and seem insufficient for navigational purposes.

Some research work has been going on in what concerns the use of artificial neural nets to process ultrasonic data. The scope of the work is twofold:

i) to exploit sensors' redundancy and be capable of making a reliable description of the environment around the vehicle. This capability is most useful in general for navigation and in particular as the basis of a position monitoring tool (see Section 7.2).

ii) To perform local navigation, that is, to be able of creating locally low level trajectories (i.e., variations to pre-defined high level planned trajectories) for obstacle avoidance, corners and doors negotiation, etc.

The approach that is being pursued builds perception maps based on generalised occupancy grids [6]. These generalised occupancy grids have been designed to take into account both the system geometrical constraints (e.g., the location of the sensors around the vehicle), and the spatial characteristics of ultrasonic sensors. Figure 17 shows the grid pattern around the mobile robot and Figure 18 shows how the grid is filled in for a particular environment and robot's position.

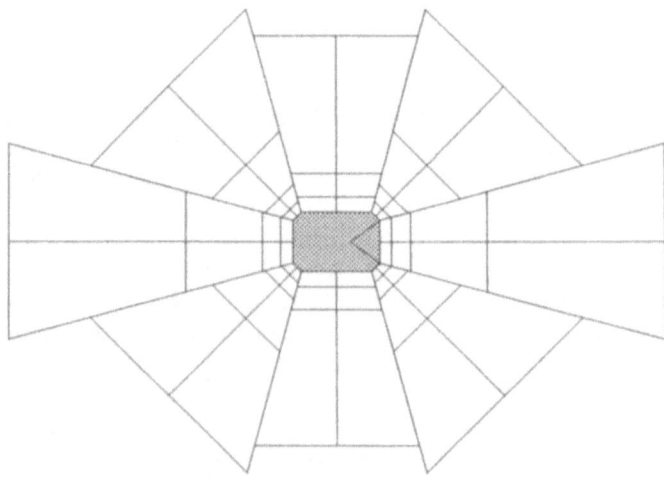

Fig. 17: Generalised occupancy grid pattern around the mobile robot.

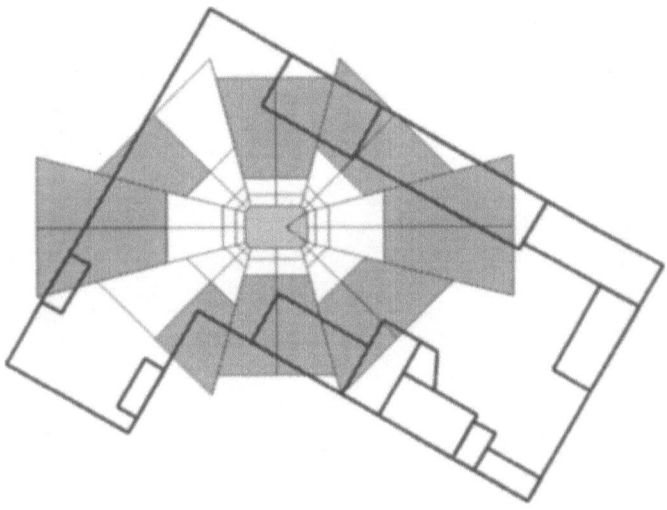

Fig. 18: Contents of the generalised occupancy grid for a particular environment and for a given robot's position.

10.2 3D Environment Modelling

The results obtained so far from the use of range information encourage the continuation of research work under this topic. Apart from improving environment authentication tools (following both the statistical and structural approaches), there is interest to investigate the construction of 3D environment models based on range data measured by the laser range finder.

Range data can be used to build 3D models of the environment, i.e., a description of its topology and its contents [17]. Such a description would be most useful for both verification and navigation purposes. A question which is continuously being raised is how sure is the *a priori* description of the environment (e.g., its CAD model)? In other words, what is the use of planning for navigation (including position monitoring and calibration) and environment authentication if reality is not mapped into a model? Such environment modelling tool could constitute an answer to this problem. Indeed, there have been requests for the use of such tool for the verification of plant design information.

Some work regarding the use of range images for 3D representation has already been developed. Figure 19 shows the 3D representation of an office scene. Several details are worth notice, namely: the office and cupboard door handles, the thermostat on the wall, the five drawer handles. Figure 20 shows the same image overlaid with a representation of the office furniture. Figure 21 shows a 3D image of the office, which results from the integration of several range images, each one taken from a different "viewing" point. The density of measurements (i.e., white points) is not uniform. Regions with larger densities are those that were scanned more than once. For a given scan (i.e., angular) resolution, multiple scans originate a more refined representation of the environment.

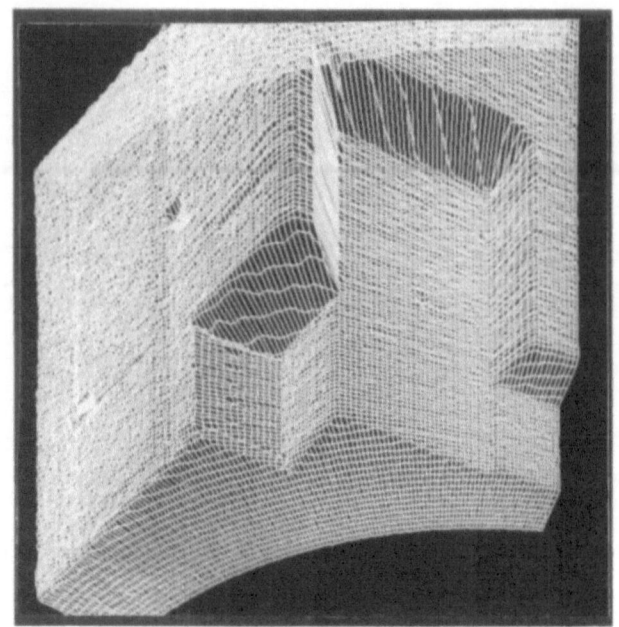

Fig. 19: 3D representation of an office scene.

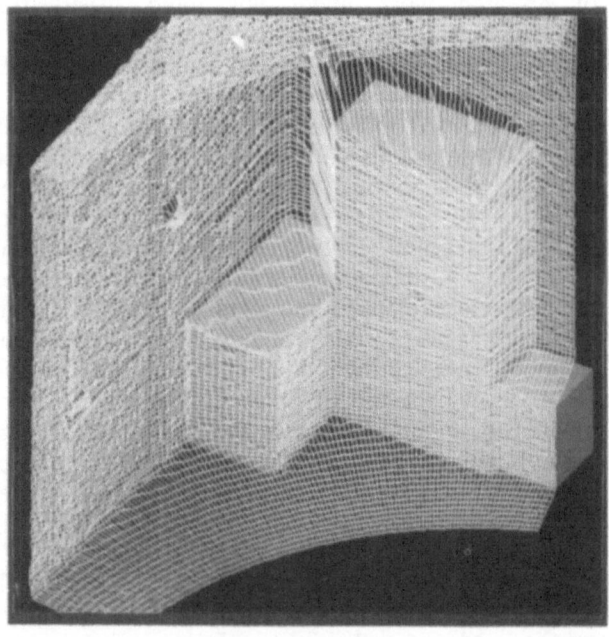

Fig. 20: 3D representation of an office scene with furniture overlaid.

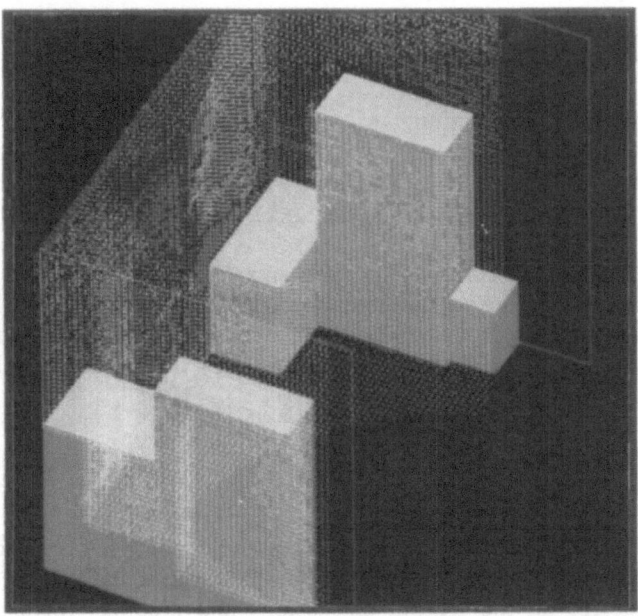

Fig. 21: Integration of several range images into a single 3D representation.

11. Conclusions

The heterogeneous computer architecture proved to be a good and economic means of achieving good performance both at the human computer interface, and at the sensor interface levels. The UNIX graphical workstation was mainly used for application oriented tasks (including user's interface aspects), and the two real-time computers for sensor oriented tasks.

The division of the project in independent, self-contained tasks, provided a good means for system's modularity. Another advantage is the possibility of porting specific software for other applications with no major efforts.

The three criteria that were specified (see Section 4) were quite helpful in distributing the different tasks amid the available processors.

The system's software and hardware architecture, and consequently the system's performance depend on the type and characteristics of the sensors. Indeed, once having set up the system's objectives and functions, system's design becomes highly dependent on the sensing devices, their precision, their bandwidth, weight, etc. It can be said that with different sensors, the design structure both at hardware and software levels will certainly be different.

The hierarchy constituted by navigation, position monitoring and position calibration, is an efficient answer for the robot's localisation problem, adapted to the characteristics of the existing sensors.

The precision of the laser range finder and its directionality constitutes a major asset for the whole project. Range information is most valuable for position calibration and for the authentication of environment scenes.

Range image segmentation attempts to extract spatial features from the image. Two methods were presented both looking for geometrical features, i.e., edges and planar surfaces. The algorithm for the detection of planar surfaces, though very computer intensive, produces better results than the one looking for edges.

A statistical algorithm based on range occupancy grids proved to be quite efficient in authenticating environment scenes. The algorithm shows good discriminating ability in detecting position errors, and specially in what concerns orientation errors. The useful range of grid cell sizes was found to be between 7.5 and 10 cm.

Acknowledgements

The authors wish to thank Mr. F. Sorel for all the encouragement and for having provided the conditions for the development of this project, Messrs. M. Cuypers and D. Landat for all the fruitful discussions and precious feedback and Mr. S. Colzani for his carefully taken photographs.

Gilberto Campos, Vítor Sequeira and Filipe Silva acknowledge the Commission of the European Communities - Joint Research Centre for the research grant that made possible their contribution to the project. Vítor Santos acknowledges the "Junta Nacional de Investigação Científica e Tecnológica", Portugal, for the research grant that made possible his participation in the project.

References

1. Al-Hujazi E. and Sood A., "Residual Analysis for Range Image Segmentation and Classification", NATO A.S.I. on "Active Perception and Robot Vision", 16-29 July, Maratea, Italy, 1989.
2. Campos G. and Gonçalves J.G.M., "DM90-215 Laser Range Finder and Command Software Library", JRC Technical Note No. I.92.132, December 1992.
3. Campos G. and Gonçalves J.G.M., "Segmentation of Range Images Based on the Split and Merge Paradigm", JRC Technical Note No. I.93.01, January 1993.
4. Cox I.J., "Blanche - An Experiment in Guidance and Navigation of an Autonomous Robot Vehicle", IEEE Transactions on Robotics and Automation, vol. 7, No.2, pp.193-204, April 1991.
5. "DM90-215 Laser Distance Meter User's Guide", Radartechnik & Elektrooptik Gmbh., Austria.
6. Elfes A., "Using Occupancy Grids for Mobile Robot Perception and Navigation", IEEE Computer, Vol. 22, No. 6, pp.46-57, June 1989.

7. Gonçalves J.G.M., Veiga P.M., Campos G. and Sorel F., "Computer Aided Tele-Operation Applied to Safeguards", Proc. of the 31 st. Annual Meeting of the INMM (Institute for Nuclear Materials Management), pp. 566-571, Los Angeles, July 1990.
8. Jain R.C. and Jain A.K. (Eds.), "Analysis and Interpretation of Range Images", Springer Verlag, 1990.
9. Landat D., "LASCO: Plan d'Ensemble", JRC internal report, June 1989.
10. Ley J., "SAOV: Stockage Avancé d'Objets de Valeur", JRC internal report, July 1986.
11. Santos V., "Classified Bibliography for Neural Navigation Project", JRC Technical Note No. I.92.100, September 1992.
12. Sequeira V. and Gonçalves J.G.M., "Permanent Auto-Focus System for Safeguards Applications using Digital Image Processing and Analysis, JRC Technical Note No. I.92.10, 1992.
13. Sequeira V. and Gonçalves J.G.M., "A Graphical Environment for the Navigation of a Mobile Robot", Proc. of Compugraphics'92, pp. 336-345, Lisbon 1992.
14. Silva F. and Gonçalves J.G.M., "A Graphical Tool for the Tele-Operation of a Manipulator Arm", Proc. of Compugraphics'92, pp. 346-354, Lisbon 1992.
15. Taylor R.W., Savini M. and Reeves A.P., "Fast Segmentation of Range Imagery into Planar Regions", Computer Vision Graphics and Image Processing, vol. 45, pp.42-60, 1989.
16. Robosoft - "Local Area Ultrasonic Network: User's Manual", June 1992.
17 Roth-Tabak Y. and Jain R.C., "Building an Environment Model using Depth Information", IEEE Computer, vol.22, No. 6, pp.85-90, June 1989.
18. Veiga P., Gonçalves J.G.M. and Sorel F., "A Man-Machine Interface for a Tele-Operated Vehicle", Proc. of IEEE Melecon'91, pp. 927-930, Ljubljana, May 1991.

Data Fusion for Environmental Monitoring

Thies Wittig
Atlas Elektronik GmbH, Bremen

Hao-Nhien Pham
Laboratoire d'Informatique Avancée de Compiègne
Lyonnaise des Eaux Dumez, Compiègne

Abstract

Environmental Monitoring is a very general term, comprising the whole spectrum of sensors and sensing - from local sensors to remote sensing from satellites. The term also implies a variety of sensor data processing aspects - from simple recording and report-generation to highly complex propagation models and prediction. This paper describes the aim of an ESPRIT project[1] that focuses on the monitoring of river and ground water on the basis of distributed sets of sensors.

1. Introduction

Environmental Monitoring is of concern for a large variety of different users and institutions and - in the end - can not be treated as a local problem. The fusion of sensor data, the assessment of the observed situations and the subsequent interpretation constitute a hierarchy of fusion/interpretation levels with increasing abstraction, both in terms of results as well as in terms of user concern.

Figure 1 depicts such hierarchy, where at the lowest level we find the sensor data fusion for a confined area of the environment that mainly caters for the local site engineer, who is concerned with the state of a river section or a specific waste dump.

The next level is the Area Monitoring, e.g. a large section of a river that runs through a densely populated area. At this level, monitoring is less concentrating on single sensor data but more on the fusion results of the level below. On the

---

[1]ESPRIT Project 6757 Environmental Monitoring Systems - EMS. The partners are Atlas Elektronik, Lyonnaise des Eaux Dumez, URA CNRS Compiègne and Technische Universität München.

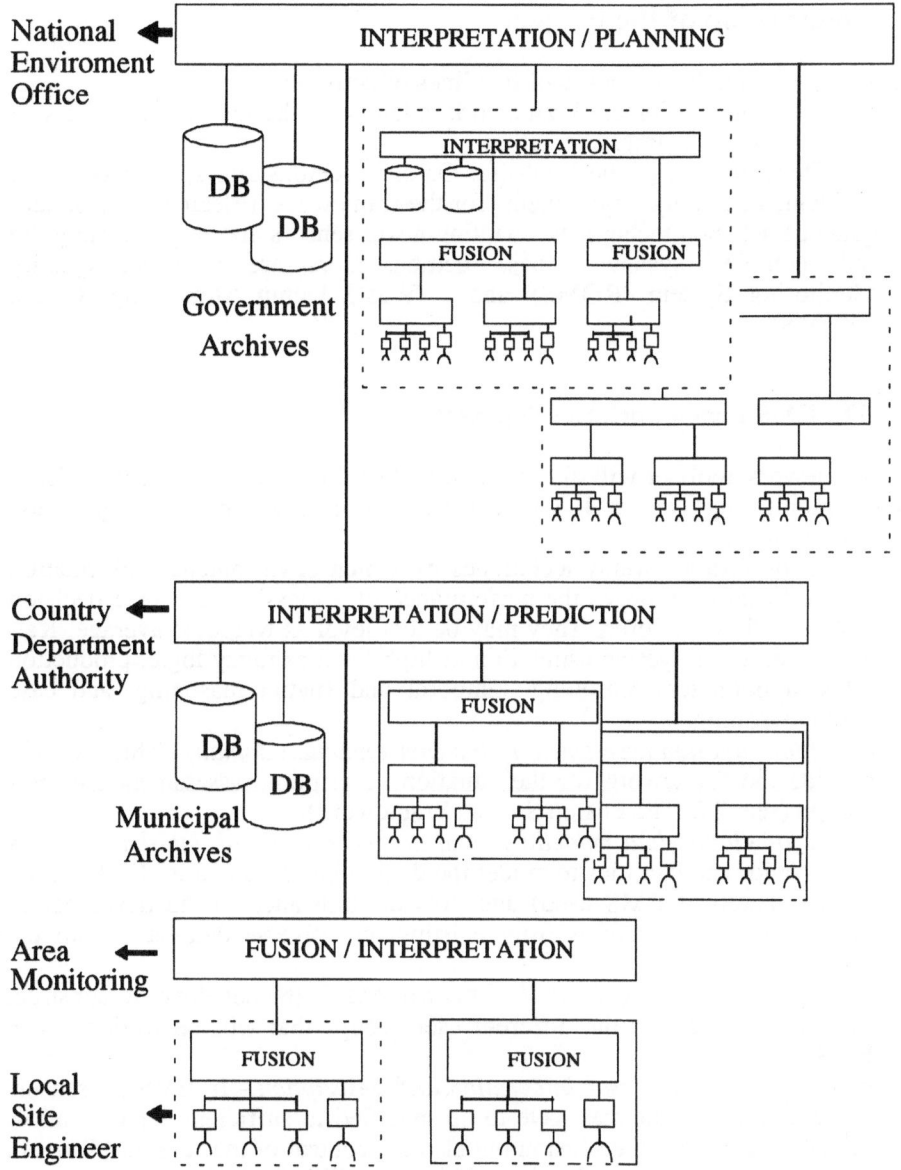

Fig. 1: General Environmental Monitoring Hierarchy

next level, the Municipal or Department Level, not only the findings of the environmental state as monitored are considered, but recorded information from archives is taken into account. This level is concerned with prediction of the evolving state of the environment required for planning purposes. Finally, for example a National Environmental Office would concentrate all the information from the municipal level in order to set up long term plans for environmental protection and the appropriate legislation.

2. Description of the Project

The project is mainly composed of two lines of activity:
- The Development of the Framework, addressing the architecture, the system design and its implementation
- The Development of the Real-life, On-site Demonstrators, addressing the problems of issuing requirements for the framework, research of most adequate data-fusion techniques, adapting novel sensors and implementing the full, complex systems. These demonstrators are a 'River Quality Monitoring System (RQMS)' and a 'Waste Dump Monitoring System (WDMS)'.

2.1 The EMS Framework Development

The framework tool set will allow a fast realisation of an intelligent environmental monitoring system. It will consist of four tools that the application developer can use:
- *a tool for fusion* - Many techniques by which environmental information can be fused to improve the performance of a classification or a decision system will be explored. They may be low level as weighted average, Kalman filter and Bayesian estimation or high level as fuzzy logic, production rules, hypothetical reasoning, temporal and spatial reasoning, and case based reasoning.
- *a tool for situation assessment in the environmental domain* - This tool will provide the framework for the situation description, assessment and prediction (see: 2.2. The Fusion System Framework).
- *a tool for global data modelling and access to external data-bases* - This will provide the facilities to model the data to be contained in EMS (based on the specific DBMS used) and provide assistance to the developer in translating data structures from existing and diverse data-bases into the EMS data-base.
- *a tool for constructing the HCI* - This will enable the developer to construct suitable interfaces to the different user groups that will be working with EMS.
- *a generic tool for sensor integration and description* - In EMS all sensors and external data-sources have to be specified in sufficient detail to allow not only the appropriate fusion but also the control of the sensors (different modes, maintenance, calibration). This tool will help the developer to enter all relevant data.

2.2 The Fusion System Framework

The term Fusion, as used in Environmental Monitoring, can be defined as the combining of data and information from multiple sources to provide a comprehensive picture of the environment and its development trends. The key system requirements are:

- *Completeness* - At least all critical environmental information must be made available for the fusion process.
- *Timeliness* - The Situation Description must be available when it is needed by Application Authorities to prevent critical environmental situations.
- *Accuracy* - Environmental information from all available sources must be fused such that uncertainties associated with a single source are eliminated to the greatest possible extent.
- *Consistency* - Environmental information from all available sources must be fused so that an overall consistent description can be generated for different Application Authorities in different levels of detail.

The above key requirements have to be viewed under a number of constraints:
- *Types of Information and Information Sources* - The appropriate information sources for the environmental monitoring system are on-line (real-time) sensor data channels, off-line sensor data, like chemical analysis in laboratory, reports and studies, natural language information of environmental observations, data bases, geographic information systems, archives with recorded time dependent data.
- *Types of Uncertainty* - The handling of uncertainty is considered to be a basic innovative problem to Environmental Monitoring. This uncertainty arises because of multiple reasons: Noise Corrupted Sensor Information, Conflicting Multi-Source Information, Low-Confidence Information, or Non-Unique Information. A significant variety of uncertainty factors will be associated with environmental information. Therefore, EMS will incorporate an evaluation scheme in the fusion process that is capable of measuring the level of uncertainty at hand.
- *Real Time Requirements* - Processes having an impact on environmental conditions under consideration are assumed to have characteristic time scales of changes that must be taken into account by the process of Situation Description. Moreover, environmental information arriving the monitoring system may contain exceptional events that drastically change the described situation and hence the basis for directives to actions.

The overall fusion process can be structured in three layers: situation description, situation assessment, and situation prediction.

The first layer has to provide the basic map of all the information from available sources with the attached degree of certainty or credibility. The second layer, the situation assessment, has the goal of assessing the described situation according to specific objectives, for example to support adequate courses of action by Application Authorities. This assessment has to take into account certainty and credibility factors, knowledge about causal dependencies, and, if available, cases from the past. The third layer, the situation prediction, provides a forecast of the most probable development trend of the environmental situation. This prediction makes use of past situations and model-based or expert-based causal and temporal correlation.

These three layers, together with the basic system components, form the EMS Logical Model, as shown in figure 2.

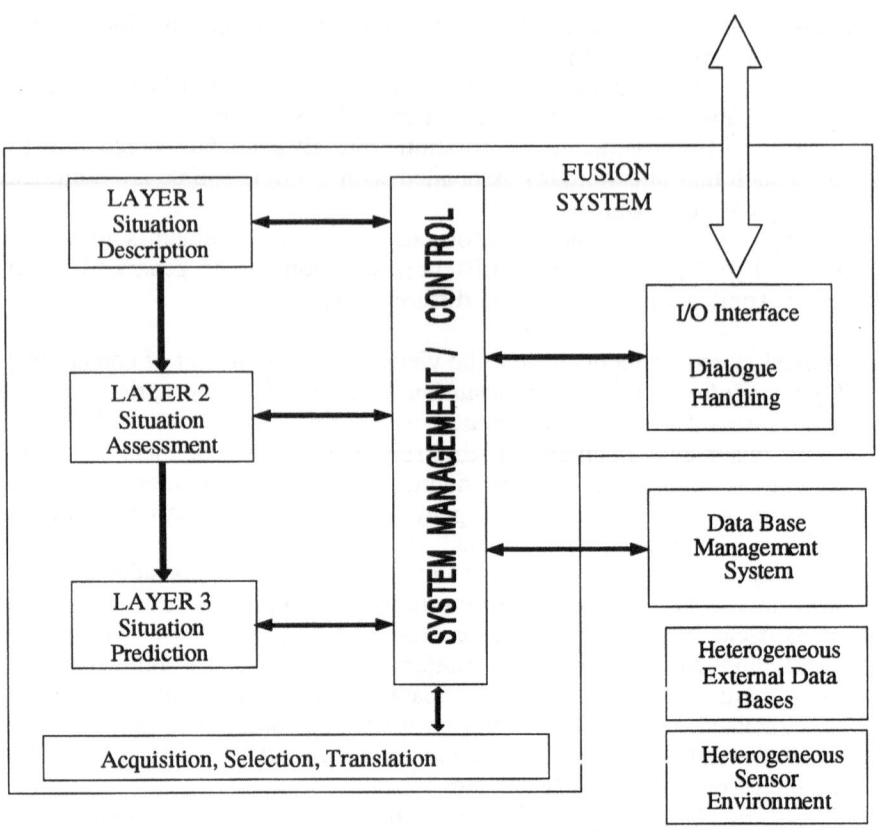

Fig. 2: EMS Logical Model

3. Modelling and Architecture

This section presents the main modelling aspects related to the physical and logical sensors, together with the general architecture of the EMS system.

3.1 The Physical Sensor Model

To describe a sensor as a black box, it is not sufficient to only define its output data format but further attributes have to be included that may have an impact on the way these data are interpreted. Some sensors have different operational and measurement modes. The characteristics of these must be described and their control sequences be given. All these definitions will be included in the *Physical Sensor Model*. This model will be used by a fusion process to access sensor data and process them according to their individual characteristics.

Although this is called *physical* model, it is important to note that it also covers external data-base entries that are needed for a fusion process. The

principal information to be contained in such a sensor model is shown in the
following table:

Interface/Transmission:
        Data Access Path: { e.g. path name for external DBs }
        Transmission-Noise: { relevant for analogue sensors }

Sensor
        Sensor-ID: T3467
        Type: 'W&G xyz1'
        Parameter   Temperature
        Meas. Interval   60 sec
        Last Measurement: <date-time>
        Data: single
        Range: $(-2.0°, 20.0°)$
        Sensitivity: .1 °
        Reliability: 1
        Accuracy: $(((-2.0°,7.0°),.04),((7.1°,20.0°),.1)))$
        Stability: (Running_days * 0.001)
        Calibration Interval: 2 days
        Maintenance Interval: (127, 92)
        Operating Modes: normal I calibration I maintenance
        Location:

Sensor-Control:
        Calibration Command:
        Mode Control:

Data-Format:
        (real-value, Meas_Mode, Oper_Mode, Stability)

For a number of chemical sensors the maintenance date - given here as the day-
number and the year - is very important. Their stability deteriorates quite
rapidly, sometimes within a week. Thus a stability function taking into account
the running_days after the last maintenance is needed for the fusion process.
The following diagram shows how a fusion process can utilise the sensor
models when accessing different physical sensors.

      The important point to note is the distinction between sensor related
processing that is contained in the sensors and the fusion related processing.
The algorithms for the sensor (pre-)processing are private to the sensor, in fact
in most cases this is proprietary code not visible to the user.

3.2 Physical Sensor Interfacing

Sensors are currently equipped with a large variety of physical interfaces
(analogue, standard digital) which EMS must be able to connect to. Therefore,
the physical sensor interface on the EMS side must be configurable, e.g. in

252

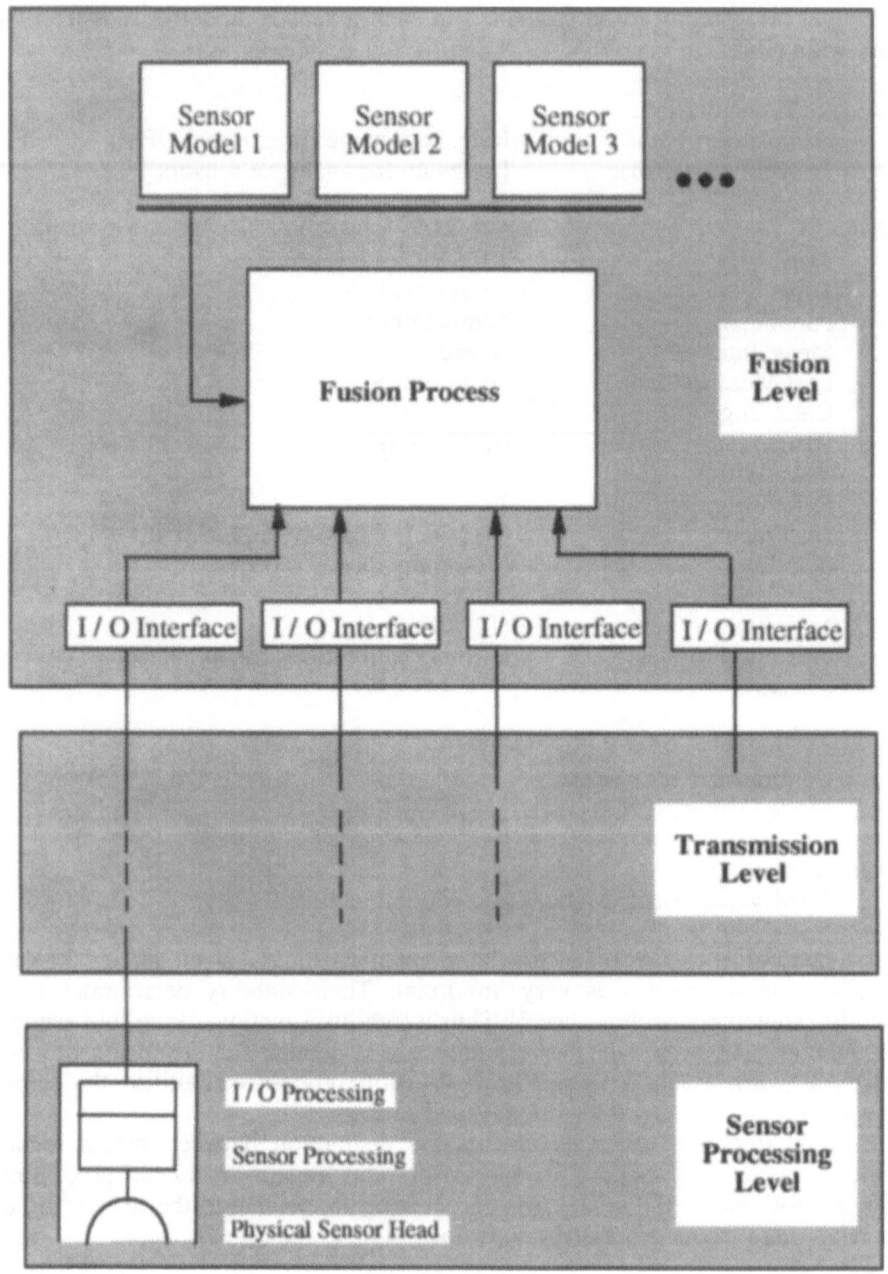

Fig. 3: Fusion and Sensor Processing

terms of "signal to measurement value relations" or in terms of "signal to connector pin references". A corresponding tool supporting this kind of physical sensor interfacing includes:

- a generalised part for access/transformation of sensor signals via a standard protocol/format (= transformation table applicable for all sensors in the environmental monitoring area), and
- a sensor specific part for access/transformation of sensor signals supplementing the generalised part according to the sensor-specific requirements.

3.3 The Logical Sensor Model

In many applications it could be tedious for the fusion process to deal explicitly with each and every single data-source. Quite often it is more natural to deal with a physical sensor ensemble that represents different sensors that support each other and belong logically together. Consequently, any fusion method related to such a sensor group belongs to the ensemble as well. The advantage of this kind of approach is a layered fusion process with clearly distinguishable levels of abstraction. Essentially, different sensors are combined to a Logical Sensor that can then be treated quite similarly to the Physical Sensor model described before: it has a number of characteristics that can be described in definition tables; it has internal processing functions, i.e. the related fusion process, as a complex physical sensor has internal pre-processing; and it has a number of operational modes that can be controlled from the outside. The way logical sensor models can be linked is shown in figure 4. A more detailed structure of the Logical Sensor is shown in figure 5. In addition to the definitions given above for the Physical Sensor model, for the Logical Sensor model additional and more detailed information has to be specified.

Definition of Operational Modes: This unit provides the functions to control the mode of operation and to report on it on request. Therefore, this unit of the Logical Sensor includes:

- an internal control module for different modes of operation such as calibration (periodical, spontaneously excited ...), maintenance, or normal. Furthermore it controls sampling rates, averaging time intervals, ...
- a mode-oriented data communication between the Logical Sensor units, and
- the provision of a Logical Sensor report containing information on events such as last calibration time, last maintenance time, exceptional events, or actual sampling rate.

Sensor Output Validation: This unit provides the functions for sensor output validation. Under operational aspects validation procedures are strongly coupled with actual, operational conditions calling for the incorporation of application specific knowledge. Therefore, corresponding tools support the implementation of validation procedures. In general terms this unit of the Local Logical Sensor includes:

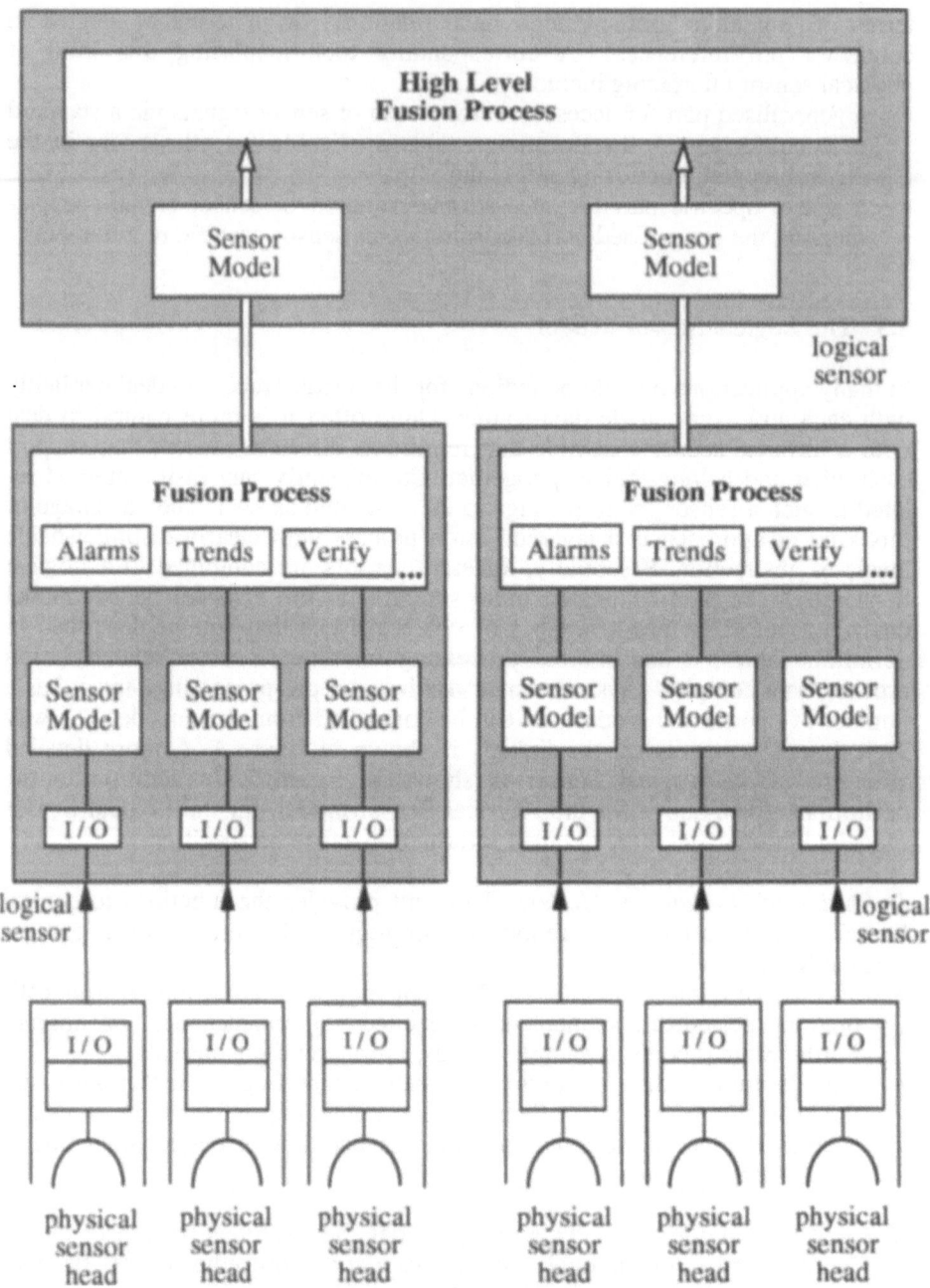

Fig. 4: Linking Local Logical Sensors

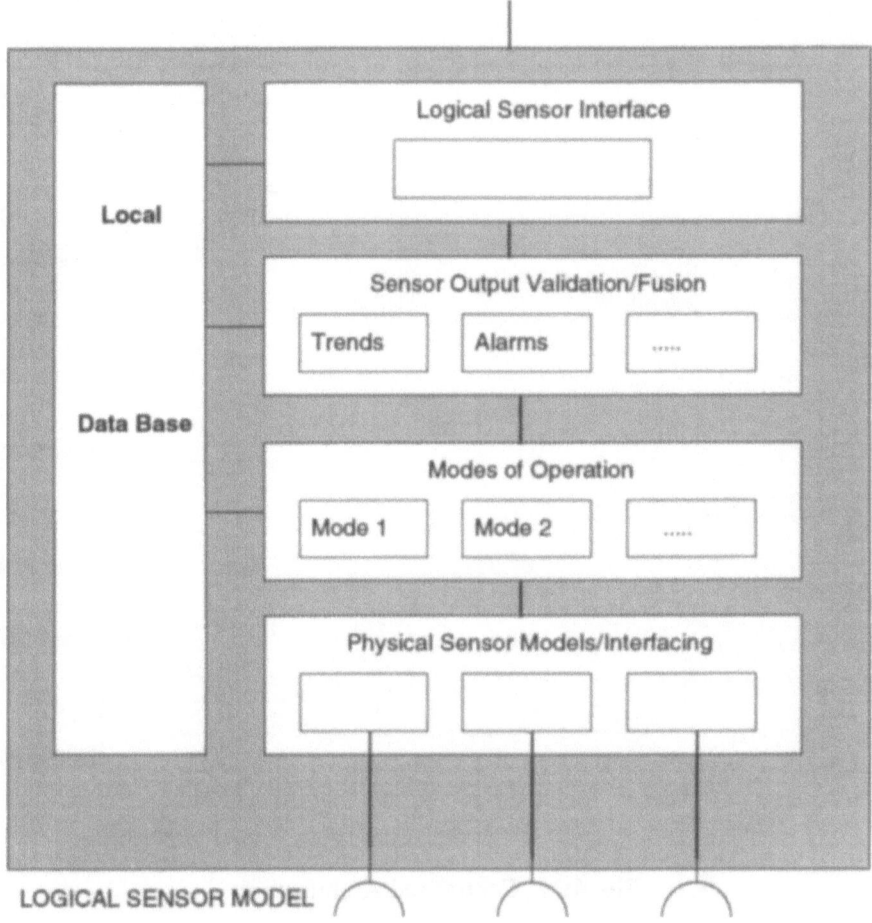

Fig. 5: The Logical Sensor Model

- Estimations of "true" measurement values based on actual calibration curve vs. target curve, sensor-inherent drift and time distance from last calibration, discontinuity in measurement values in connection with re-calibration.
- Determination of confidence values for measurement value estimated on the basis of:
  - error-relevant indicators/constraints (sensor-inherent noise characteristics, maintenance status, external measurement conditions as established by means of supporting sensors), and
  - historical cases (observed slopes of curves from measurement values, observed mean squared deviations).
- Threshold alarm generation.

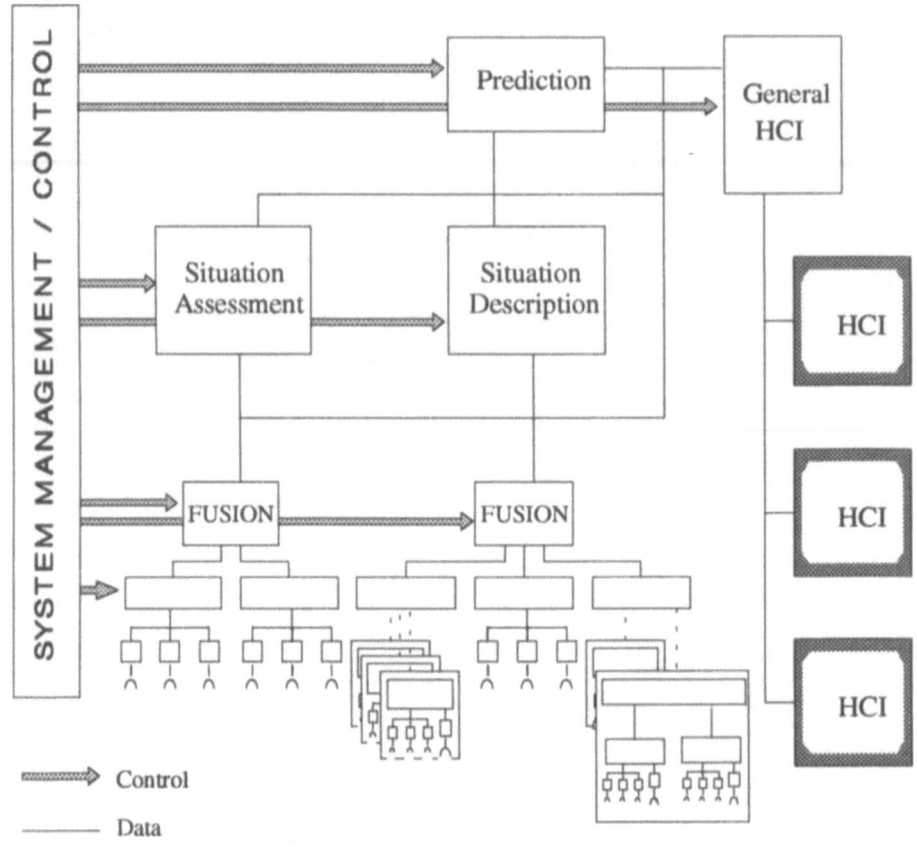

Fig. 6: A Centralised Architecture

- Determination of statistical parameters associated with measurement values (e.g. time averages, fluctuations), with weightings according to confidence values, or with adjustable time granularity.
- Calculation of pollution components by means of combination/correlation of directly measured parameters or by utilising a priori knowledge about material compositions.

3.4 General Architecture Options

The general Logical Model describes the functional modules and their dependencies. The next step is to turn this logical model into an architecture. In principle there are two quite different architecture options: the *Centralised* and the *Distributed*.

In the Centralised Architecture only one System Management and Control component exist. Likewise the modules that provide Situation Description, Assessment and Prediction only exist once. Depending on the level of abstraction or reasoning, they take care of all data from all sensors. The

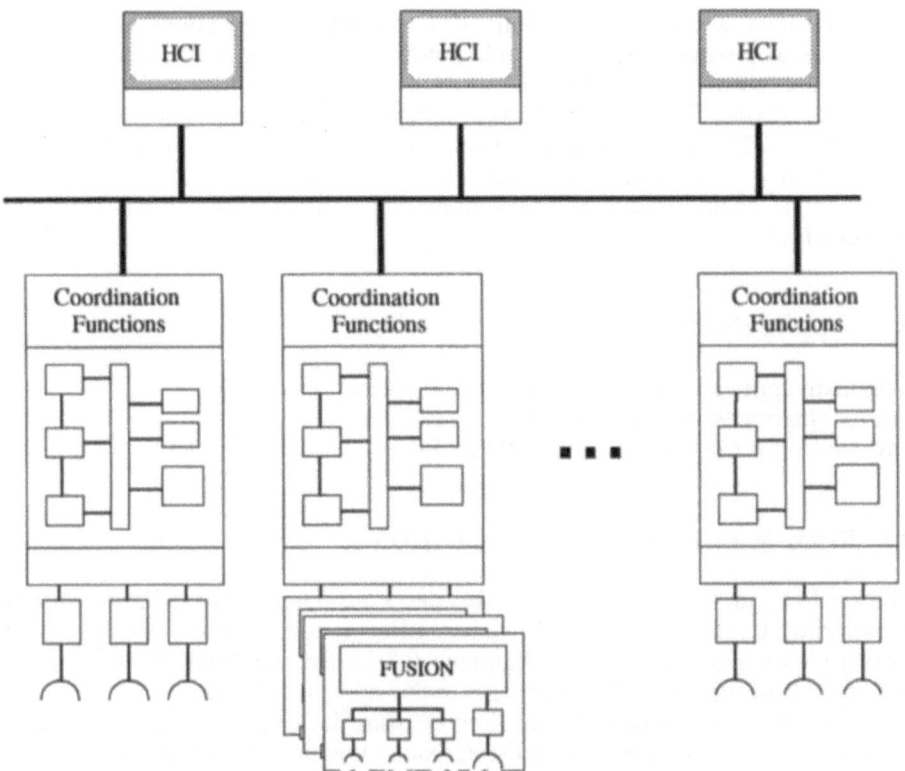

Fig. 7: A Distributed and Decentralised Architecture

resulting information is supplied to a General HCI component. Attached to this are individual HCIs for specific user groups that are able to filter, transform and present information stored in the General HCI in a fashion best suitable for that specific user group (see figure 6).

This system approach appears to be best suited for locally confined applications with a not too large number of sensors. Likewise it seems appropriate for applications where sensors can be easily grouped into sets and where the inter-dependencies between these sets are straightforward. However, a problem may be encountered if, e.g. at a later stage, the application has to be extended to take into account a much larger area (referring to the General Monitoring Hierarchy presented above, this is a horizontal extension). Maintaining the same singular modules and extending their processing and data-handling capabilities may quickly lead to fairly complex systems. Splitting these modules into two or three, however, will be difficult without major changes to the general architecture, i.e. the System Management and Control Module.

From this situation to the second architecture option is but a small step: Instead of splitting single modules, the entire system would be replicated as shown in figure 7. The important feature of this second option is that there is no 'higher level' system management module.

The idea is that with the help of 'Co-ordinating Functions' attached to each of these systems they will be able not only to communicate but also to co-operate, i.e. jointly solve problems.

The second difference to the first option is that here no general (central) HCI exists. The HCIs can profit from this architecture by behaving more independently and having individual access to specific systems to obtain information, again driven by the requirements of the user group they are intended for.

4. Applications

Two main applications will demonstrate he project result:
- River Quality Monitoring System (RQMS), and
- Waste Dump Monitoring System (WDMS).

4.1 River Quality Monitoring System (RQMS)

The application provided by Lyonnaise des Eaux-Dumez is on River Quality Monitoring. Lyonnaise des Eaux-Dumez (LED) is responsible through its sub-sidiary Compagnie des Eaux de Banlieue (CEB) for supplying drinking water to hundreds of thousands of inhabitants in the suburbs of Paris.

The water resource is taken directly from the river Seine and treated in treatment plants. Moreover, the West Parisian Direction of LED produces water by using underground water resources situated near the river Seine. Due to increasing urbanisation of this area, needs in drinking water have increased while underground resources have decreased. This situation leads to a lack in the natural supply of aquifers so that raw water is pumped from the river Seine to filter through the ground into the underground aquifers. These facts show clearly that the water supplier needs a continuous analysis of the river quality states. This analysis leads to adapt the water treatment plant functioning according to the river water quality.The RQMS will be at CEB for monitoring both the drinking water supply and the river quality. This monitoring will be based on the following existing data sources:
- water quality monitoring stations,
- laboratories,
- public authorities,
- drinking water supplier, and
- national weather services.

Informal information coming from other users will be added to this list. These data can be classified in three types:
- *Real-time data* - coming from monitoring stations: pH, temperature, dissolved oxygen, conductivity, turbidity, ammonium, dissolved hydrocarbons, UV and visible absorbency, ...
  The importance of the real-time measures for the overall quality situation assessment is due to the fact that these data are essential for alert and pollu-

tion crisis triggering. In that context, sensor improvements in the sense of increasing their reliability or sensitivity are of main interest. The role of maintenance in monitoring stations is a crucial aspect of reliability.

- *Normal delayed time data* - coming from laboratory analysis. Weekly: Organoleptic parameters: aspect, turbidity, physico-chemical parameter, hydrotimetic level, calcium, magnesium, chlorides, sulfatides, aluminium, dissolved oxygen, saturation percentage, ...
  Quarterly and yearly: analysis following EEC standard for 120 parameters.
- *Exceptional delayed time data* - provided in case of pollution. Depending on the type of pollution, data describing the nature of pollution, quantities and dates are provided to the drinking water supplier by firemen, navigation authorities, other drinking water suppliers in case of chemical pollution, by waste water treatment plants in case of waste water disposals, by sewage network manager in case of rainfall disposals.

From the point of view of drinking water supplier, the river quality should be evaluated by identifying several states:

- The present normal state, which takes into account more recent data and which allows an overall judgement on the overall river water quality. There is no need for the operators to take any particular action.
- The present exceptional state, for instance during a pollution crisis. The more recent data should be used, but the acquisition frequency should be increased in order to provide more precise information.
- The close past state, which corresponds to several hours or days. This state is related to an a posteriori analysis of crisis phenomena or to quality parameters evolution according to forecasts made by numerical models.
- The far past state, required to a posteriori analysis for determining the consequences of long term river management policy. This analysis requires information on water parameters' quality on several months or years.
- The near future state, used for determining, in a forecasting approach, the short-dated consequences of exceptional phenomena (e.g. pollution, lowest water level, floods). These forecasts will be provided by numerical models or by past data analysis. This last approach consists of finding parameters' evolution in past situations close to the present one.
- The far future state. This state will be defined for evaluation of consequences of environmental policy or new investments. These forecasts require use of past and present data.

Application users should have access to information according to the kind of responsibility they have in system management. Each type of potential user will require different information depending on his need and his behaviour.

4.2 Waste Dump Monitoring System (WDMS)

The Waste Dump Monitoring System is intended to provide particularly safe and environment-friendly management of waste dumps. It is based on comprehensive, systematic acquisition and analysis of parameters relating to

monitoring and operation. The system information can be used during the future lifetime of the dump as a significant aid to the execution of the following tasks:
- Implementation of the pay-as-you-pollute principle (tracing emissions back to the source).
- Potential regression (the waste dump as a store of raw materials).
- Removal of particular substances after danger has been detected.

The typical end-user requirements are:
- *Dump Transparency* - At the entrance area of a dump, the main data of each incoming load (quantity, constituents, identity of the company delivering the load) are recorded in a data-base. On the basis of these data, a suitable location within the dump is determined. Again this location is entered into a data-base so that at any time all important parameters of the dump (substances and their location) are available.
- *Removal of Dangerous Material* - In the event of unexpected emission, e.g. contamination of percolating water and ground water, it is easy to locate the possible sources of the pollutants. The necessary investigations can be clearly targeted, and the subsequent removal of dangerous material can thus be confined to a limited area. This leads to a considerable reduction in the cost of removing such material.
- *Systematic Monitoring of Emission* - By systematic monitoring of emission, the emission behaviour of the dump is known at all times. If suitable sensors are used, data can be acquired about contamination of surface water and ground water. Furthermore, automatic warnings are issued by the system if relevant parameters exceed given thresholds. In addition, the emission of dust can be significantly reduced by analytical feedback between tipping control and emission behaviour.

The data to be handled in WDMS are:
- Data determined in the entrance area of the dump, such as date and time, weight, contents of the load, or identity of the company delivering the load.
- Sensor information, such as physical parameters, meteorological parameters, pollutants (sensors in drilled holes, in drainage systems, etc.) to determine the contamination of the surface water and ground water.
- Waste dump input data-Databases. This serves to determine the presence of potential pollutants contained in individual materials arriving at the dump.

The dump model combines the following sub-models:
- Pollutant Transport Model. The knowledge of hydro-geologists and dump experts has to be integrated into an transport model to determine the propagation behaviour of individual constituent substances in percolating water, surface water and ground water.
- Dump Model. All the dump activities have to be incorporated into a three dimensional dynamic model of the dump where the information based on the individual tipping operation and the declaration of the incoming loads are combined to a general dump model.

The dump model can be used to process the sensor signals and database contents numerically in order to obtain a dynamic, three-dimensional map of the dump. In addition to standard output that help to organise the dump or to graphically visualise statistical trends, display facilities can be created so that the answers to the questions "Where is waste A situated?" or "What is present at location X?" can be presented as clearly as possible.

Acknowledgement

The work described here is the result of a group of researchers and developers who have come together not only to set up a framework for Environmental Monitoring but also to develop and implement it over the next three years. Among those engaged in this work we want to acknowledge the contribution to those aspects described here by M. Clement and A. Santoni from Lyonnaise des Eaux Dumez; by K. Henschel, N. Börsken, W. Ernst and W. Jung from Atlas Elektronik; by S. Canu and T. Denoeux from URA CNRS Compiègne and from T. Baumann from Technische Universität München.

References

1. Afsarmanesh H., McLeod D., "The 3DIS: An Extensible Object-Oriented Information Management Enviroment", ACM Transactions on Information System, Vol. 7, No 4, pp. 339-377, Oct. 1989.
2. Arnold U., Datta B., Hänscheid P., "Intelligent Geographic Information Systems (IGIS) and Surface Water Modelling", Proc. IAHS Congress 89, Symp. 3, pp. 407-416, Baltimore May 1989.
3. Ayache N., Faugeras O.D., "Building, Registrating and Fusing Noisy Visual Maps", Research Rapport No 596, INRIA-Rocquencourt, Dec. 1986.
4. Decker K.S., "Distributed Problem Solving Techniques: a Survey", IEEE SMC, Vol 17, No 5, 1987.
5. Hanson A.R., Riseman E.M., Williams T.D., "Sensor and Information Fusion from Knowledge-based Constraints", in Proc. SPIE, Vol. 931, pp. 186-196, Sensor Fusion, Weaver C.W., Orlando Florida, April 1988.
6. Henderson T., Weitz E., Hansen C., "Multisensor Knowledge Systems, Robotic Research, Vol. 7, No 6, pp. 114-137, MIT Press 1987.
7. Henderson T., Shilcrat E., "Logical Sensor SystemS", J. Robotic Systems, Vol. 1, No 2, pp. 169-193, 1984.
8. Huntsberger T.L., Jayaramamurthy S.N., "A Framework for Multi-Sensor Fusion in Presence of Uncertainty", in Proc. Workshop Spatial Reasoning and Multi-Sensor Fusion, St. Charles Illinois, pp. 345-350, Oct. 1987.
9. Larzewski H., "Datenbanken in der Umwelttechnik", UTB-Berichte, No 15, Berlin, June 1989.
10. Slade S., "Case-Based Reasoning: A Research Paradigm", AI Magazine, Vol. 12, No 1, 1991.
11. Wittig T., "ARCHON: - an Architecture for Multi-Agent Systems", Ellis Horwood, 1992.

Concluding Remarks:
Advanced European Research on Data Fusion

K.C. Varghese
Commission of the European Communities,
DG XIII, Brussels

S. Pfleger
Technical University of Munich

The workshop "Data Fusion Applications" presented recent research directions within the ESPRIT framework. The impact of the emerging data fusion technology on the recent European research is expressed by a number of 22 technical and overview papers reflecting the expertise of 69 experts in computer applications. Advanced research results on data fusion have been here discussed together with practical solutions in multi-sensor data fusion in a wide area of applications: medical diagnosis and patient monitoring, real-time expert systems, robotics, monitoring and control, marine protection, surveillance and safety in the public transportation, image processing and scene interpretation, and environment monitoring.

Computer systems operating in public areas, and integrating visual sensors with various electrical and chemical sensors are required to continue correct operation in the presence of several classes of operational faults. The ability of combining several types of sensorial data obtained from many sources of information - data fusion - has the advantage of providing robust problem solving in the presence of missing or faulty sensorial data, due to information redundancy and diversity.

In medical applications, complementary sensorial observations need to be integrated from different medical viewpoints in order to permit a correct problem detection and solving. This is a typical data fusion task applied to *medical diagnoses*. An inovative data fusion application in the neuro-radiological diagnosis of the cerebral blood vessels is presented by Bahner et al. in [1]. The Information provided by Magnetic Resonance Angiography and Digital Subtraction Angiography is here integrated (e.g. position, orientation, width and branching) and allows a real visualiation (i.e. a three dimensional view) of the re-constructed cerebral blood vessels surrounding an aneurysm. In this way the neuro-radiologist is supported by the computer during the diagnosis process. Several classes of medical expert systems have been developed in order to assist the experts in the decision processes of medical diagnosis and therapy.

Monitoring of the artificial respiration support, suggested by Gärtner et al. in [2], is one of the critical applications in which the fusion of diverse human specific data is performed by computer in order to improve the *decision process* in patient surveillance systems.

Fuzzy logic is an appropriate tool to model and describe human experience and uncertain computer knowledge in the decision-making process. The diverse data types (e.g. force-torque sensor data and mini-camera visual data discussed by Zhang and Raczkowsky in [8]), are weighted evidential observations, and influence the fuzzy inference. These observations are fused and used as control data for a robot gripper with the aim of performing a correct operation. An inovative real-time expert system is presented by Aquilar-Crespo et al. in [9], and uses fuzzy logic techniques and data fusion for providing assistance to running petrochemical processes in a large production plant.

The new technology of reactive actions in *robot task planning* is based on fusion of multisensorial data (e.g. as implemented by Grunwald in [10]), and intends to increase the robustness of the robot actions in an unpredictable environment. The fusion of odometric data and laser range scanner is investigated by Ekman et al. in [11] with the intention of providing a global view of the environment for semi-autonomous robots. A mobile robot is presented by Goncalves et al. in [20], which is capable of performing *remote surveillance* tasks (e.g. visual inspection, reading of electronic seals, remote measurement) inside of fissile material storage areas.

The fusion of visual and non-visual observations improves the *quality of the image processing and interpretation tasks*. The support of image segmentation by the fusion of intensity data with the range data of a microwave radar is addressed by Siebert et al. in [12]. The support of the matching process, as presented by Moneta et al. in [13], is based on vectorial representations of both structural similarities of the compared objects and the information about the reliability of the available descriptions. A new data integration technique in the process of tracking edges is presented by Tistarelli in [17]. Distributed knowledge-based support is suggested by Regazzoni in [14] for the integration of several image processing modules. A new technique is pesented by Peri et al. in [15] for the evaluation of the crowding level in complex scenes, based on the spatial fusion of multisensor visual information. The fusion of two views using object motion is suggested by Hogg and Baumberg in [19], and its suitability to the surveillance of pedestrian scenes is discussed.

Data integration in *public transport* surveillance applications is the aim of the ESPRIT project DIMUS. The prototype system developed during the last two years is presented by Benvenuto et al. in [6]. An inovative approach to reflex-based fusion of visual data is presented by Bozzoli et al. in [7]. Fusion of non-visual sensorial data (e.g. of movement detectors, tactile arrays using force and position sensing resistors as presented by Hagen and Witte in [18], and beam breaker) is suggested by Pfleger and Milano in [16] as a complementary approach to image processing for the detection of persons in dangerous situations. Dynamic evaluation of the sensorial alarm reliability is here performed with the intention of avoiding false alarms.

Information fusion in *monitoring* applications using the Category Model is discussed by Steinke in [3].

Monitoring and control of vessel traffic using multisensor data fusion is presented by Stefanelli [4] in the ESPRIT project TRACS.

Environmental monitoring is a critical research topic, comprising the whole spectrum of sensors and sensing classes, from local sensors to remote sensing by satellites. The ESPRIT project AZZURRO investigates the protection of marine environment. The project results are here presented by Ghelfo [5] together with the fusion aspects of environmental data, collected by two different sensors: an active sensor such as Lidar Time Resolved Laser Fluorosensor and a passive multispectral scanner (Daedalus). The ESPRIT project EMS, as presented by Wittig and Pham in [21], focuses on the monitoring of river and ground water based on the fusion of distributed sensorial observations.

Is the problem of fusing and interpreting sensor data solved ? Not really! This book reflects the state of the art in data fusion applications. It is hoped that the practical results obtained in some areas can be re-used in other areas. However, past experience recommends caution, and future research work is needed for validating and improving the efficiency of the existing data fusion techniques.

References

1.   Bahner M., J. Dick, B. Kardatzki, H. Ruder, M. Schmidt, A. Steitz, C. Betram, D. Hentschel, T. Hildebrand, E. Hundt, R. Kutka, S. Stier, G. Gerig, T. Koller, O. Kübler, G. Szekely, "Combining Two Imaging Modalities for Neuroradiological Diagnosis: 3D Representation of Cerebral Blood Vessels", pp.1-16 (in this volume).

2.   Gärtner K. , S. Fuchs, H. Jauch "Hybrid Inference Components for Monitoring of Artificial Respiration" pp.17-26, (in this volume).

3.   Steinke W. "Information Fusion in Monitoring Applications using the Category Model" pp.27-37, (in this volume).

4.   Stefanelli L. "A flexible Real-Time System for Vessel Traffic Control and Monitoring", pp.38- 43, (in this volume)..

5.   Ghelfo S. "ESPRIT Project AZZURRO: Data Fusion for Marine Protection", pp.44-49, (in this volume).

6.   Benvenuto F. , M. Ferrettino, M. Pasquali, F. Perotti, P. Verrecchia

      "ESPRIT Project DIMUS: Data Integration in Multisensor Systems", pp.50-70, (in this volume).

7.   Bozzoli A. , M. Rossi, R. Barbò, B. Caprile, G. Carlevaro "A Reflex-based Approach to Fusion of Visual Data" pp.61-70, (in this volume).

8.   Zhang J., J. Raczkowsky, "Sensor Fusion in a Peg-In-Hole Operation with a Fuzzy Control Approach", pp.71-78, (in this volume).

9. Aguilar-Crespo J.A., J. M. Domínguez, E. de Pablo, X. Alamán
    "Fuzzy Logic Techniques for Sensor Fusion in
    Real-Time Expert Systems, pp.79-86, (in this volume).

10. Grunwald G. "Task-Directed Sensor-Planning" pp.87-101,
    (in this volume).

11. Ekman A., D. Strömberg, "Incremental Map-making in Indoor
    Environments", pp.102-114, (in this volume).

12. Siebert A., J. Ostertag, B. Radig, M. Rozman, J. Detlefsen, J. Bernasch,
    "Image Segmentation Improvement with a 3-D Microwave Radar",
    pp.115-122, (in this volume).

13. Moneta C., G. Vernazza, R. Zunino, "A Vectorial Definition of Conceptual
    Distance for Prototype Acquisition and Refinement", pp.123-132,
    (in this volume).

14. Regazzoni C., "Distributed Knowledge-based Systems for Integration
    of Image Processing Modules", pp.133-154, (in this volume).

15. Peri, C. Regazzoni, A. Tesei, G. Vernazza, "Spatial Fusion of
    Multisensor Visual Information for Crowding Evaluation", pp.155-171,
    (in this volume).

16. Pfleger S., A. Milano, "Multisensor Fusion in Underground
    Stations", pp.172-182, (in this volume).

17. Tistarelli M., "On Tracking Edges", pp. 183-195, (in this volume).

18. Hagen J., M. Witte, "Force and Position Sensing Resistors: An Emerging
    Technology", pp.196- 203, (in this volume).

19. Hogg D., A. Baumberg, "Fusing Two Views using Object Motion"
    pp.204-213, (in this volume).

20. Goncalves J. G. M. , G. Campos, V. Santos, V. Sequeira, F. Silva, "Mobile
    Robotics for the Surveillance of Fissible Materials Storage Areas: Sensors
    and Data Fusion", pp.214-245, (in this volume).

21. Wittig T., H. N. Pham, "Data Fusion for Environmental Monitoring"
    pp.246-261, (in this volume).

Contributors